S0-CWR-876

FOOD SERVICE MANAGEMENT

Helen Delfakis, RD
Teacher/Director
Hospitality Careers Program
Santa Rita High School
Tucson, Arizona

Nancy Loman Scanlon, CFE
Admissions Representative
Johnson & Wales University
Providence, Rhode Island

Janis B. Van Buren, PHD, CHE
Assistant Professor
Vocational Education Division
School of Education
Purdue University
West Lafayette, Indiana

ON10AA
PUBLISHED BY
SOUTH-WESTERN PUBLISHING CO.
CINCINNATI, OH DALLAS, TX

Production Editor: Joseph P. Powell III
Designer: Craig Ramsdell
Associate Photo Editor/Stylist: Linda Ellis
Marketing Manager: Donald H. Fox

Content Consultants:

Dr. Gwen Frazier
Vocational Consultant
Consumer Homemaking
Bureau of Vocational-Technical Education
Concord, New Hampshire

Maurine R. Humphris
Formerly, Consumer and Home Economics Specialist
Utah State Board for Vocational Education
Salt Lake City, Utah

Diane Stringer
Assistant State Supervisor
Home Economics
Division of Vocational and Technical Education
Jackson, Mississippi

ISBN: 0-538-60692-4

Library of Congress Catalog Card Number: 90-61647

1 2 3 4 5 6 7 8 9 0 K 0 9 8 7 6 5 4 3 2 1

Printed in the United States of America

CONTENTS

PREFACE

The fastest-growing service industry in the United States today is the food service industry. The restaurant business, which creates over 200,000 new jobs every year, represents a major segment of this industry. The food service industry also includes businesses involved in food manufacturing and distribution, facilities and equipment design, food sanitation control, food marketing and writing, computer systems management, training, and health and nutrition concerns. You will find information on the food service industry's multifaceted career opportunities in this textbook.

We, the authors of *Food Service Management,* recognize the need for interesting and challenging material designed to prepare students for the successful management of food service operations. We hope that you will be able to use many of the skills developed while studying this textbook to find new or better jobs in the food service industry. We also hope that you will be able to establish career goals for which you can further your education in the specialized programs offered at vocational and technical schools and at colleges and universities throughout the United States.

Food Service Management Skills

The successful management of food service operations combines a knowledge of food production and service as well as business functions. In *Food Service Management,* you will learn about both these aspects of food service management throughout the history of food service. You will read about the responsibilities of many types of food service employees, including dining room managers, servers, and bus persons. To enhance your knowledge of service skills, you will be asked to demonstrate a variety of place settings and different ways to use napkins as decorative table accessories. Additional service skills, such as order taking and the correct service of food, are outlined. The importance of good customer relations to successful dining room management is discussed as well.

Menu planning, an essential food service management skill, receives detailed treatment in this textbook. Menu planning requires an awareness of the benefits of different food products. This knowledge is important for creating healthful and well-balanced menus. Purchasing food products and maintaining them in a sanitary and safe environment is another major responsibility of food service managers. You will learn about the illnesses that result from poor sanitation techniques and how to prevent them with proper food-handling methods.

The business skills involved with food service management include learning how to correctly calculate selling prices to produce a profit. To develop this skill, you will study recipe cards and cost cards. These cards help food service businesses create quality products consistently without exceeding the costs that management has determined will produce a profit. You will also identify the basic record-keeping functions that are the backbone of successful restaurant management and examine how food service computer systems can assist with record keeping.

Features of This Textbook

Food Service Management offers some unique opportunities for you to explore different aspects of the food service industry. Each chapter begins with vocabulary words, which are highlighted and defined in the chapter. Clearly stated objectives are also set out for you at the start of each chapter. These objectives are directly tied to the questions and activities at the end of the chapter.

Most chapters include a Career Profile that examines the career of a person actively involved in a food-service-related field or describes a job associated with the topic of the chapter. Interviews with a food writer, a research and development chef, a tabletop designer, and an entrepreneur offer insight into just a few of many possible career choices.

Nutrition Notes in many chapters highlight a variety of health concerns and the different ways that the food service industry is responding to them. From the basic principles of balanced food plans to healthful eating programs linking area hospitals and local restaurants, these notes will expose you to the application of good nutrition principles.

Math Screens in a number of chapters allow you to carry out a variety of math functions directly related to food service management. These screens give you the opportunity to apply your math skills to the daily operations of a food service business, such as calculating the cost of labor and establishing selling prices for food items. In addition, Alcohol Awareness features throughout the textbook offer guidelines for serving alcoholic beverages responsibly.

Activities at the end of each chapter allow you to apply what you have read in a variety of challenging and creative situations. Developing a marketing campaign, doing research in community restaurants, creating a restaurant promotion, and preparing a number of delicious recipes are a sampling of the many activities offered in this textbook.

Features of the Student Supplement

The Student Supplement is designed to provide you with hands-on experience and help you develop marketable working skills by exploring various activities within the food service industry. These activities may be completed by yourself or with a group of students. They include designing a commercial kitchen layout, planning a menu, completing mock purchasing forms, and practicing the different methods and techniques used in cooking by preparing a variety of fun and deliciously different recipes. In addition, you will practice filling out a job application and creating a résumé, and you will become oriented to the world of work by learning how to prepare for, secure, and hold a job. In addition, you can test your knowledge of the textbook's content by completing the matching and vocabulary activities provided.

Features of the Teacher's Resource Guide

The Teacher's Resource Guide provides teachers with the tools they need to make optimum use of textbook and workbook materials, and it assists teachers in the instructional design of the course. The Teacher's Resource Guide is organized by chapter and contains chapter outlines, suggestions for audiovisual materials, transparency masters, a test bank, and answer keys for all activities, quizzes, and tests.

Your Future in Food Service

The food service industry is currently the number one employer in the United States, with over 20,000 managerial positions and $230 billion in sales annually. Projections suggest that, by the year 2000, one of every ten employees in the United States will be working in a hospitality- or food-service-related position.

This textbook is designed to prepare you for a career in the fast-growing food service industry in order to meet its increasing demand for qualified employees and to provide you with more job options. Career opportunities are available in hotels, restaurants, supermarkets, recreation facilities, catering businesses, clubs, employee feeding services, schools, hospitals, colleges, universities, the military, transportation companies, day-care organizations, and major food corporations, from entry-level to top management positions.

High school graduates who have completed courses in restaurant management or food service may apply directly for positions. Or, they may choose to pursue a postsecondary degree in hotel and restaurant administration; restaurant, hotel, and institutional management; nutrition and food science; dietetics; or culinary arts. Programs for these degrees are offered by many colleges and universities.

About the Authors

Helen Delfakis teaches food service management and nutrition at the secondary and postsecondary levels. She is also the director of the Hospitality Careers Program at Santa Rita High School in Tucson, Arizona. This program is linked with the Southern Arizona Innkeepers/Restaurant Association in a partnership that has facilitated career explorations and job shadowing in the hotel industry and in restaurant operations. Ms. Delfakis earned a master's degree in dietetics from the University of Arizona. She is a registered dietitian, a certified teacher, and an active member of the Southern Arizona District Dietetic Association, the American Dietetics Association, the National Education Association, and the National Restaurant Association. As a teacher and dietitian, Ms. Delfakis has developed numerous standardized recipes and nutrition education materials for various groups. In addition, she has many years of restaurant service and management experience, starting at the age of 14 in a family-owned restaurant.

Nancy Loman Scanlon is an admissions representative and instructor for Johnson & Wales University. She earned a bachelor's degree from the University of Connecticut and her certified food service executives certification from the International Food Service Executives Association. Prior to joining Johnson & Wales University, Ms. Scanlon was with Hilton Hotels Corporation in the area of food and beverage management. As a teacher of restaurant management practices on the college level, she has seen high school student's interest in the food service industry develop into career goals and professional achievements. Visiting and lecturing in high school classrooms has given her an insight into the needs of students and teachers for information and materials about food service management. As a consultant and member of the Pennsylvania Restaurant Association, she works actively with industry professionals to promote hospitality and food service education. Ms. Scanlon is the author of the college textbook *Marketing by Menu*. Her professional memberships include the National Restaurant Association, the International Food Service Executives, and the Roundtable for Women. The Roundtable for Women named Ms.

Scanlon a 1990 Pacesetter, a national award in recognition of her contributions to the food service industry.

Janis B. Van Buren is an assistant professor in the Vocational Education Division, School of Education, Purdue University. She received her doctorate from Iowa State University in home economics education with an emphasis on food and nutrition. Dr. Van Buren has taught food and nutrition courses at Northeast Missouri State University and has been responsible for food service in congregate meal sites in Michigan. She has edited two food and nutrition texts for South-Western Publishing Co., has given numerous nutrition and health presentations, and has served as a nutrition educator. As a gender-equity specialist, she is aware of the differential treatment females and males receive in the food service industry as well as other vocational- and technical-education-related careers. Her professional memberships include the American Home Economics Association, the American Education Research Association (AERA), the American Vocational Association, and their state affiliates. In 1990, she received the national AERA Women in Education Award for curriculum development. Dr. Van Buren is a certified home economist.

Helen Delfakis, RD
Nancy Loman Scanlon, CFE
Janis B. Van Buren, PhD, CHE

PART ONE

PERSPECTIVES IN RESTAURANT MANAGEMENT

1
RESTAURANT HISTORY

VOCABULARY

vomitorium
volume feeding
restaurant
truffles
commis
apprentice
tavern
inn
ordinary
table d'hôte
à la carte
American plan
European plan
delicatessen
cuisine

OBJECTIVES

After studying this chapter you should be able to do the following:

- Outline the basic history of restaurant development in the United States
- Discuss the various ways that meals have been served during different historical periods, from ancient Greece to present-day America
- Explain the two methods of pricing restaurant meals
- Explain the differences between the European and American plans
- Describe the development of the fast-food industry in the United States

Food has been a major factor in world history. People, moving away from their homes in order to trade, explore, and travel, needed cooked foods. Originally, these travelers prepared food in temporary facilities. The increasing demand for ready-to-eat food, however, created the need for what has become known as the restaurant. The purpose of this chapter is to review the history of food service over the past two thousand years as well as the history of restaurant development since its beginnings in early nineteenth-century Paris.

FOOD SERVICE ORIGINS

Food has been a central part of life throughout human history. The "Golden Arches" and the elegant hotel dining rooms of the 1990s have stone caves and open-pit fires at their roots. Berries, raw meats, and roots were the first staples of the human diet. When agriculture developed, the cultivation of grains and vegetables supplemented dietary intake. Cooking techniques evolved for preparing meats, poultry, and fish. Cheeses and other milk products were added as many new methods for cooking protein-rich foods emerged. Each geographical region of the world slowly developed its own distinctive cuisine in response to available foods and agricultural techniques.

ILLUSTRATION 1–1 The "Golden Arches" and the elegant hotel dining rooms of the 1990s have stone caves and open-pit fires at their roots.

As societies in Europe, Asia, India, and Africa created distinct and increasingly complex cultures, the role of food preparation and service gradually changed. People became more sure of their food sources and were able to concentrate on developing industries and governments, which in turn further ensured the food supply. Food became a pleasure more than a burden.

Ancient Greece and Rome

Greece and Rome set the standards for food service in the ancient world with sophisticated food preparation and service. In Greece, upper-class dining consisted of a series of small dishes or appetizers served with a variety of wines and a final course of sweets, cheeses, and fruits. Three couches were placed in a U shape and diners reclined, supporting themselves with one arm. They ate from a central table with the other arm while dancers, musicians, and comics provided entertainment.

Rome developed an even more elaborate arrangement for large festivals and banquets. The **vomitorium** was an infamous feature of Roman architecture that provided guests a facility where they could vomit what they already had eaten so that they could return to the table and gorge themselves again.

The Middle Ages

By 1310 European food service had advanced to the point where King Richard II of England could order the ingredients listed in Figure 1–1 for a banquet feeding ten thousand guests. This example of volume feeding was certainly not a simple undertaking. **Volume feeding** means providing food for large numbers of people. To simplify the volume-feeding process, huge amounts of food were set out on a table at one time.

The Seventeenth and Eighteenth Centuries

When people traveled away from their homes during the Renaissance and eighteenth century, some type of organized commercial shelter provided food service, although it was very simple. The average person had few opportunities to eat outside the home. Street vendors in the public markets sold cooked foods for at-home consumption. Operators of stalls and shops provided seasonal take-out fare.

The first restaurant was opened in Paris in 1794. The word **restaurant** was derived from the French word *restaurer,* which means "to restore" or "to refresh." This first restaurant served a simple meal of truffles, cheese fondue, and soups. (**Truffles** are edible, potato-shaped fungi that grow underground. They are considered a great delicacy.) After the first restaurant had laid a successful foundation, other

FIGURE 1–1 Banquet for Ten Thousand

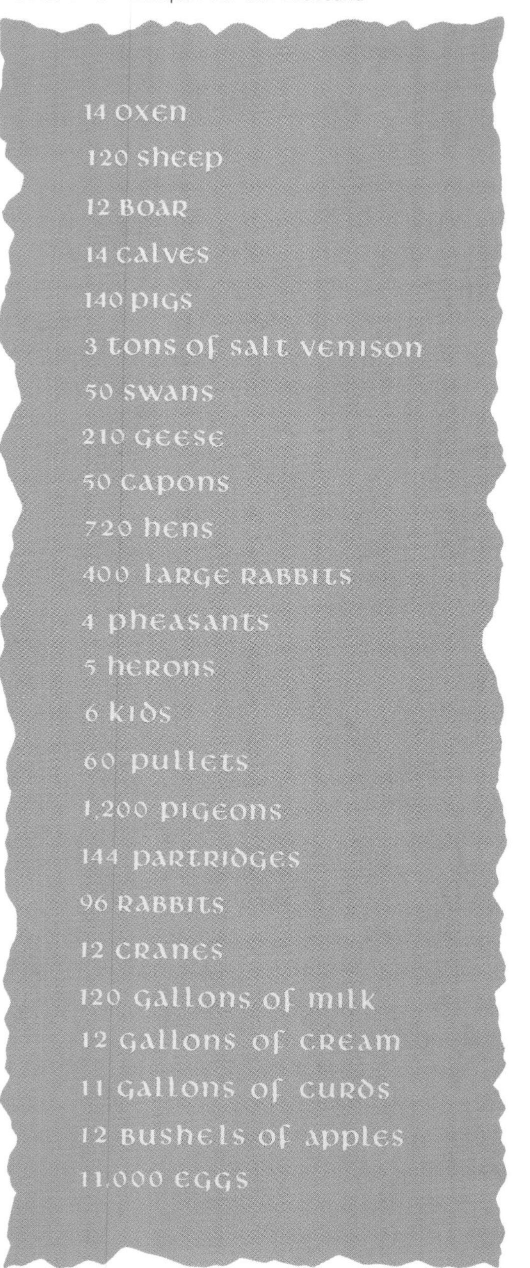

restaurants quickly developed in Paris and London.

Auguste Escoffier's Culinary Revolution

One of the most notable names in commercial restaurant history is Auguste Escoffier, who was born in 1846. After serving as a ***commis,*** or kitchen steward, he was apprenticed to chefs in France. An **apprentice** is a person who is learning by practical experience under a skilled worker in that trade. As an apprentice, Escoffier was put to work in the kitchen. He first scrubbed vegetables and cleaned. Then he learned the cooking skills required to be a chef. He went on to become a master chef, revolutionizing French cuisine and developing an organized system of commercial kitchen management. He took the complicated Medieval-style menu used in nineteenth-century France and reorganized it into the structure followed today. His menu for a dinner organized the meal into seven courses, as shown in Figure 1–2. Escoffier's revolutionary ideas also simplified the

FIGURE 1–2 Escoffier Menu Outline

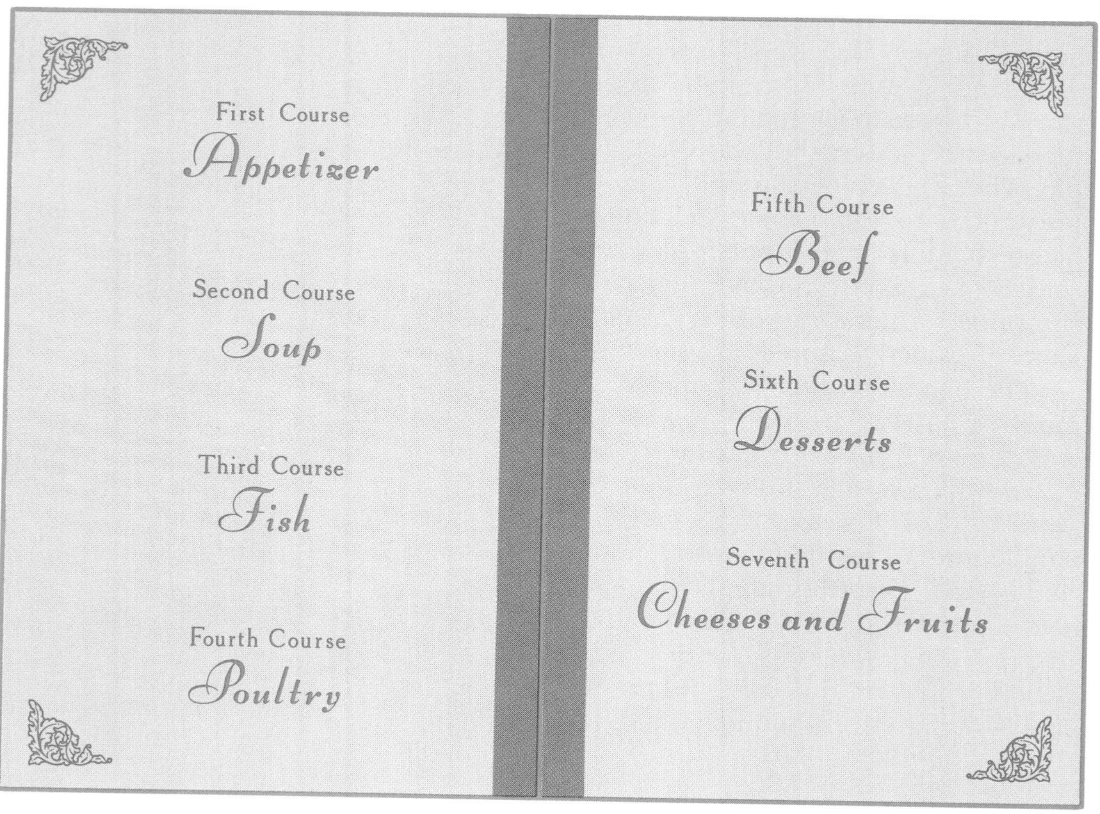

NUTRITION NOTE HEALTHFUL DINING

Healthful dining and nutritional awareness are as important today as when Auguste Escoffier created the seven-course-meal pattern. This new type of menu, along with revised recipes, helped to reduce the cholesterol, fat, and calorie counts of many restaurant meals. Today the American Cancer Society and the American Heart Association have joined with hospitals and restaurants to create programs (such as the Dining With Heart program) that enable the average person to choose healthy meals in public restaurants. Hospital dietitians have helped chefs at participating restaurants rewrite some of their recipes to produce healthier, more nutritious meals. Through these and similar programs, the food service industry today is working with other organizations to help Americans eat better.

menu and the presentation of foods. His concern for healthier dining changed the recipes for many dishes and reduced the butter, fat, and cream content.

EARLY AMERICAN HOSPITALITY

American restaurant history had its beginnings in local inns and taverns located on the stagecoach route that ran along the eastern coast of the American colonies. As early as 1634, Cole's in Boston, Massachusetts, offered food and lodging. In 1691 travelers found Hall's further south in Salem, New Jersey. Clark's Tavern and Inn opened in Philadelphia in 1693. Better road conditions increased the amount of travel during the eighteenth century. More and more inns opened, joining the hundreds of ordinaries and taverns that provided basic food and lodging services.

Hospitality terminology in prerevolutionary America defined a **tavern** as a drinking place that served spirits. An **inn** served meals as well as spirits and alcoholic beverages. Inns generally provided lodging as well. In New England an **ordinary** was established in each town to provide travelers with a place to spend the night. Owners of taverns and inns advertised by hanging signs outside their doors.

Hotels

The number of American hotels and their dining rooms grew slowly as the country itself began to expand. It was not until 1794, for example, that the

ILLUSTRATION 1–2 Owners of taverns and inns advertised by hanging signs outside their doors.

City Hotel opened in New York City. As pioneers began settling the Midwest and Far West, they left behind the modern hotels of Boston and New York. When the English author Charles Dickens traveled by stagecoach in 1842 to Upper Sandusky, Ohio, he found himself in a log cabin papered with old newspapers.

The fast-moving Americans, however, wanted their towns and cities to move along with them, and hotels began springing up throughout the American West. The International Hotel, for example, opened in Virginia City, Nevada, in 1875. It had an excellent wine cellar and cuisine to match, thus gaining a firm place in the lore of luxury western hotels.

America's hotels were its palaces, the settings for its fine public dining rooms. By the mid-1860s, every major city boasted its own grand hotel, the most famous of which are described in the following sections.

The Waldorf-Astoria, New York City

William Waldorf Astor opened the Waldorf-Astoria on Fourth and Fifth avenues in 1893. The Waldorf offered two of the finest restaurants in the country, Peacock Alley and Oscar's. The present hotel, owned by the Hilton Hotel Corporation, still provides some of the highest-quality restaurant service in America today.

The Palace Hotel, San Francisco

The Palace Hotel is now called the Sheraton Palace. It offered elaborate menus in its dining room.

The Willard Hotel, Denver

The Willard Hotel in Denver offered a wide variety of native game and beef on its seasonal menu. These traditional American dishes featured wild game and fowl, including buffalo, grilled antelope, venison, elk, mutton, plover, wild dove, quail, partridge, ruffled grouse, teal, mallard, and sage hen. The Willard House was the first public restaurant in the West open to unescorted women.

The Parker House, Boston

In nineteenth-century Boston, great strides were made in restaurant service. Although restaurants at that time served daily meals to the public, the meals were scheduled for very specific times. Guests were offered a **table d'hôte** menu, which meant that the entire meal was included in one price. In 1855 Harvey Parker opened a Boston hotel, The Parker House, that served meals at irregular hours. He presented an **à la carte** menu, or a variety of food items from which to choose. A separate price was charged for each item. This innovation in service and menu format laid the groundwork for modern restaurant menus.

Hotel Menu Pricing

Because hotel dining rooms were among the few, if not the only, restaurants in cities, they offered a choice of payment plans. The first choice was the **American plan,** in which all three meals were included in the room rate. The second, called the **European plan,** required guests to be billed separately for their meals.

Riverboats

Not all American hotels and restaurants were on solid ground. The Mississippi and Ohio rivers supported a

ILLUSTRATION 1–3 William Waldorf Astor opened the Waldorf-Astoria on Fourth and Fifth avenues in 1893.

ILLUSTRATION 1–4 The *Delta Queen,* out of New Orleans, Louisiana, still offers overnight trips on the Ohio River.

fleet of floating hotels and restaurants. These riverboats transported both passengers and cargo. The *Delta Queen,* out of New Orleans, Louisiana, still offers overnight trips on the Ohio River.

Railroads

In the late 1800s and early 1900s, the railroads offered complete American plan menus in their dining cars. The railroad companies even developed America's first resorts by building hotels in remote locations to serve as vacation destinations. Lake Mackinaw in Michigan, Cape May in New Jersey, Sun Valley in Idaho, and Saratoga in New York are just a few locations that began as railroad hotels and restaurants and developed into established communities.

THE TWENTIETH-CENTURY AMERICAN RESTAURANT

By 1910 the great wave of European immigration to the United States had created a huge swell in the population. Millions of immigrants moved across America bringing with them diverse food and dining traditions. From these traditions many new food service outlets emerged, including the family restaurant, the roadside diner, and the **delicatessen,** a shop where ready-to-eat foods such as cold meats, salads, and sandwiches were sold.

The Great Depression of the 1930s eliminated many medium-priced restaurants. Soup kitchens and bread lines fed the many unemployed during that time. The Second World War brought a rebirth of the American restaurant, as Americans, weary of war news and rationing, looked for entertainment and the luxury of eating out. The restaurant business was stabilized in the 1950s, and the fast-food segment of the food service industry emerged in the 1960s.

Fast-Food Restaurants

America's fast-food industry has its roots in turkey and mashed potato plates, grilled hot dogs in a sleeve, and inexpensive hamburgers and chicken. From these beginnings, Hardees, Burger King, Wendy's, Pizza Hut, Taco Bell, and countless other fast-food outlets have emerged. Most fast-food restaurants' original business was based on a single food item, but increased competition has led these establishments to diversify by serving breakfast, regional cuisines, pasta, and salads. Restaurants such as Denny's, Sizzler, Bennigan's, and Big Boy have also moved into the market, offering a variety of family-style meals and children's menus.

Howard Johnson's

The bright orange roofs of Howard Johnson's restaurants became America's first fast-food landmark. They provided food service facilities along the emerging network of American superhighways. The Howard Johnson's concept established a bridge between family dining and fast food by providing a range of food service options, from complete dinners to walk-up ice cream windows.

McDonald's

When the McDonald brothers opened their hamburger stand in San Bernadino, California, they were convinced that they could sell a hamburger and french fries for a price well within a typical family's budget. After Ray Kroc took over and McDonald's went public, hamburgers were still only 49 cents and an order of french fries cost 29 cents. The chain's original marketing theme was "Change back from your dollar."

Kentucky Fried Chicken

In Louisville, Kentucky, a gas station owner known as Colonel Sanders claimed that his southern fried chicken was the country's best. He sold it out of a storefront to a swelling crowd of customers. By 1989 Kentucky Fried Chicken was the largest fast-food chicken speciality chain in the United States.

American Cuisine

While the fast-food industry has exploded, American cuisine has quietly flourished. **Cuisine,** which is originally a French word, refers to a manner or style of cooking. Fine-dining restau-

ILLUSTRATION 1–5 When the McDonald brothers opened their hamburger stand in San Bernadino, California, they were convinced that they could sell a hamburger and french fries for a price well within a typical family's budget.

rants around the country feature various American cuisine, including Cajun cooking, Tex-Mex, and California Nouvelle.

CAREER PROFILE
ELAINE TAIT: FOOD COLUMNIST, PHILADELPHIA

By ELAINE TAIT

Ms. Tait has been writing for the *Philadelphia Inquirer* for the past 25 years and still says that "I love this job—creative, flexible, always changing, it is the ultimate job in a constantly evolving business, complete with deadlines and stress."

Her interest in food started as a 14-year-old cook at home. Food has always fascinated her, says Ms. Tait, for its colors and textures as well as for its taste.

Ms. Tait started writing accidently; she majored in home economics and fashion design, not in journalism, at Drexel University in Philadelphia, Pennsylvania. After graduation, food continued to intrigue her enough to answer a job ad from the *Philadelphia Inquirer* for someone interested in food and writing. Today Ms. Tait is one of the most widely read and respected restaurant critics and food columnists in the country.

Sensitive to the impact of her restaurant reviews, Ms. Tait feels that her responsibility as a reporter is not only to the consumer, but also to the restaurant. The restaurant must be viewed honestly, claims Ms. Tait, with a direct appraisal of its successes and its failings.

Ms. Tait not only covers restaurants but also is responsible for a food column. She feels that the current focus on nutritional awareness has had a positive impact on American society and the way in which Americans approach food.

SUMMARY

From its humble beginnings in small inns and taverns, the American food service industry has answered the growing demand for quality food and good service. Hotels, riverboats, railroads, and delicatessens have all met the public's changing needs for a variety of food outlets, from expensive restaurants serving new American cuisines to fast-food chains offering quick, inexpensive meals.

American restaurants have benefited from many past innovations, including the ancient Grecian and Roman emphasis on guest comfort, Medieval volume-feeding techniques, and the development of à la carte menus in nineteenth-century Boston. The French chef Auguste Escoffier had an especially profound effect on modern menus and food preparation.

Now is an exciting time to be a part of the American food service industry. Increased competition and a growing number of American cuisines promise many dynamic changes as the restaurant business moves into the twenty-first century.

QUESTIONS

1. Explain the origin and meaning of the word *restaurant.*
2. Outline the changes in the way meals have been served from ancient Greece to modern America.
3. How did à la carte and table d'hôte menus influence the way food items are priced on restaurant menus today?
4. What role did hotels play in the history of American restaurant development?
5. List some of the great American hotels. Which of these are still operating today?
6. Who were the founders of America's fast-food industry?
7. What influence did Howard Johnson's have on twentieth-century food service development?
8. List the two payment plans used by hotel restaurants. What distinguishes one pricing method from the other?
9. Define the word *cuisine* as it relates to American restaurant development.
10. What is an apprentice?

ACTIVITIES

Activity One Update the banquet for ten thousand menu by substituting foods that would be appropriate to American cuisine of the 1990s.

Activity Two Outline the growth of the fast-food industry in the United States. Choose one of the major fast-food restaurants and present to your class a history of the company and an evaluation of how it compares with other food service chains. Use the periodical section of your school and public libraries to find reference material.

2

FOOD SERVICE MANAGEMENT CAREERS

VOCABULARY

food service management
organizational chart
back-of-the-house positions
front-of-the-house positions
kitchen brigade
pastilliage decorations
food service marketer
purveyor
dietitian
institutional food service
sanitary engineers
health inspectors
food stylist
food writer
product developer

OBJECTIVES

After studying this chapter you should be able to do the following:

- Identify career opportunities within the food service industry
- Create an organizational chart for a typical food service organization
- Describe the front-of-the-house and back-of-the-house food service positions
- Choose a food service job that matches your personal abilities, goals, interests, and skills
- Describe types of educational programs available in the area of food service

In the 1990s approximately 70 percent of all high school students will have a food service job as their first work experience. Fast-food and family style restaurants provide many job opportunities for unskilled workers. The hourly wages and flexible schedules are perfect for students and retired adults looking for part-time and summer work. Very few students, however, continue to work in the food service industry after high school or college. Often, this is partly because they are unaware of the range of chal-

ILLUSTRATION 2–1 In the 1990s approximately 70 percent of all high school students will have a food service job as their first work experience.

lenging positions and the financial rewards available in this field.

The purpose of this chapter is to introduce you to career opportunities in the food service business. Perhaps you will find a career interest in this fast paced, creative, and dynamic industry.

THE FOOD SERVICE INDUSTRY

The food service industry is the fastest-growing service industry in the United States. Providing over 200,000 new jobs annually, it has a powerful impact on the American economy. The National Restaurant Association estimates that the average family eats two of its three daily meals outside of the home. Americans drive through, take out, or eat out for at least 15 breakfasts, lunches, or dinners each week. These meals are purchased and served in many different settings, including cafeterias, diners, snack bars, fast-food outlets, full-service dining rooms, delicatessens, airplanes, and stand-up counters. No matter what or where Americans eat, however, the food has been designed, prepared, and sold according to a carefully orchestrated pattern developed by professionals in the food service industry.

Food service management is the coordination of people, resources, products, and facilities related to the design, preparation, and presentation of food outside of the home. Food service management as a career offers a wealth of job opportunities, from food preparation and cooking in a local restaurant, to developing new food products for food manufacturers, to designing restaurant interiors and commercial kitchens.

ILLUSTRATION 2–2 Food service management is the coordination of people, resources, products, and facilities related to the design, preparation, and presentation of food outside of the home.

FOOD SERVICE POSITIONS

Restaurant operations require a team effort. In order to understand the type of positions available in the food service industry, it is helpful to look at organizational charts from restaurants and their kitchens. An **organizational chart** is an outline of all of the job categories within an organization. On such a chart, jobs are placed from top to bottom in order of decreasing responsibility.

Whether independently owned or part of a hotel or chain, most restaurants have the same basic organizational chart, as shown in Figure 2–1. The number of positions in each category depends on the size of the restaurant. Note that Figure 2–1 is broken down into two separate sections. **Back-of-the-house positions** are those jobs in all areas outside of public space, and **front-of-the-house positions** are involved with guest service.

Front-of-the-House Positions

Most restaurant-goers are familiar with the front-of-the-house workers in the food service industry. The following sections describe these jobs in more detail.

Restaurant Manager

In a freestanding restaurant, the restaurant manager assumes responsibility for the overall front-of-the-house operations (and oversees back-of-the-house operations as well). Handling guest relations, staffing, and training; determining food and beverage costs; maintaining the physical plant; and following safety and sanitation practices are the main components of this job.

Assistant Manager

This position is usually reserved for full-service restaurants. The assistant

FIGURE 2–1 Restaurant Organizational Chart

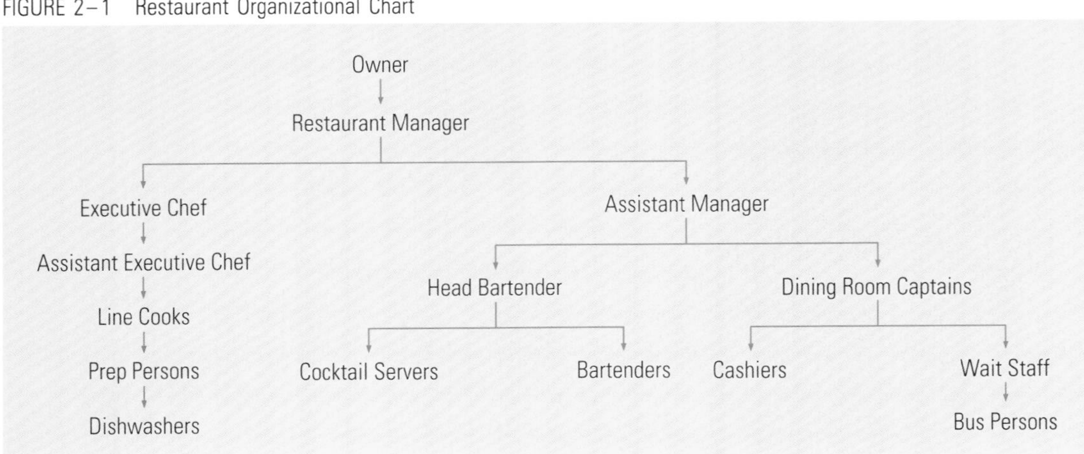

ILLUSTRATION 2–3 In a freestanding restaurant, the restaurant manager assumes responsibility for the overall front-of-the-house operations.

manager acts as a host, greeting and seating guests, and is also responsible for guest relations, the smooth functioning of the dining room, and some staff training.

Dining Room Captains

Captains are responsible for overseeing guest relations and service within different sections of a dining room. Table-side preparations and presentations are usually their duty; thus they must have a good knowledge of the wine list and menu.

Additional help in the dining room includes hosts and hostesses, who work directly under a captain, or who may substitute for captains in smaller restaurants. Cashiers handle customer payments, an important part of guest relations.

Wait Staff and Bus Persons

These employees are responsible for serving all guests in a professional, friendly manner. An awareness of changing menu items and effective service techniques are important aspects of these jobs.

Bar Staff

Bartenders work directly under a head bartender and are required to know how to mix a wide variety of both alcoholic and nonalcoholic beverages. Cocktail servers, like the wait staff, must treat guests in a courteous way.

Back-of-the-House Positions

Every kitchen includes a team of individuals who, under the direction of the executive chef, perform all the food production tasks for a restaurant. This team is called the **kitchen brigade.** The chef who oversees this team must have the management and culinary skills to keep the kitchen brigade functioning smoothly at all times.

The classical French kitchen has a formal organization that delegates specific jobs in the kitchen to cooks whose specialty is preparing certain types of food. Although American kitchens in many ways do not function like their counterparts in France, they do follow the classic *brigade de cuisine* structure. Figure 2–2 (on page 18) shows a typical organizational chart

ILLUSTRATION 2–4 Every kitchen includes a team of individuals who, under the direction of the executive chef, perform all the food production tasks for a restaurant.

for a kitchen brigade, with titles in English and French. Depending on the size of the operation and kitchen, some positions in the kitchen brigade may be eliminated.

Executive Chef

Also known as the *chef de cuisine,* the executive chef has similar job responsibilities whether working in a restaurant or a hotel. The executive chef is in charge of the entire kitchen. In addition, this chef often acts as a purchasing agent, ordering all of the food items required by a facility. A working knowledge of the ways that computers can be used in food service management has become a basic requirement for this position. Most executive chefs have also worked the majority of the other kitchen positions before attaining the highest rank.

Assistant Executive Chef

The assistant executive chef, often called the *sous chef,* is responsible for the kitchen when the executive chef is absent. This person may be called upon to assume the responsibilities of any one of the line cooks, depending on his or her individual specialty. The assistant executive chef also needs a

FIGURE 2–2 Kitchen Brigade *(Brigade de Cuisine)*

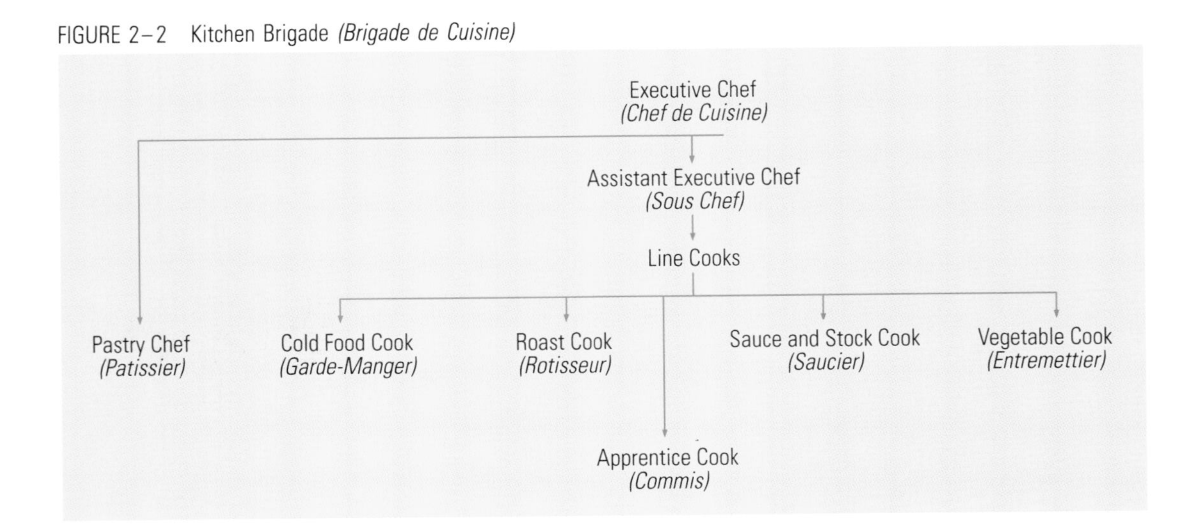

degree of management ability and knowledge of computers and software used in food service.

Cold Food Cook

The *garde-manger* is a creative position due to the amount of show work involved in buffets and special functions. This cook is responsible for attractive cold food trays, hors d'oeuvre trays, and buffet presentations, and he or she usually specializes in vegetable and ice carving.

Roast Cook

The *rotisseur* prepares all main food items consisting of meat, poultry, and fish. Depending on the food preparation styles of the restaurant, the roast cook may work primarily on the range, the grill, or the broiler. This position requires a great deal of stamina and concentration. Working conditions are often uncomfortable due to heat, and the *rotisseur* is usually under intense pressure for long periods of time.

Sauce and Stock Cook

The *saucier* is responsible for the preparation of sauces and stocks. In addition to their cooking duties, many sauce and stock cooks also serve as assistant executive chefs. This position is very specialized and requires extensive training.

Vegetable Cook

The *entremettier* prepares all vegetable items and soups. Sometimes this position involves dessert preparation as well. Like the sauce and stock cook, the vegetable cook often works on the line. This cook must spend considerable amounts of time cutting and otherwise preparing vegetables before cooking.

Pastry Chef

The *patissier* attends to all baking and pastry making. Breads, rolls, cakes, pastries, and desserts are prepared according to the needs of the establishment. The pastry chef often does

ILLUSTRATION 2–5 The *patissier* attends to all baking and pastry making.

special presentation or show pieces depending on his or her specialty. These elaborate ***pastilliage*** **decorations** are made of a hard sugar paste. The pastry chef must be a creative person with specific training in pastry arts.

Apprentice Cook

The *commis* is an apprentice cook who is training to become a chef by working as a kitchen assistant. In the United States, formal apprenticeship programs are administered by the American Culinary Federation (ACF). These programs are usually associated with established culinary programs in two- or four-year colleges around the country. The ultimate goal of many culinary students is to become a master chef, the highest standing in the culinary profession.

Apprentices usually work as short-order cooks, prep persons, and dishwashers, some of the lowest-paying jobs in the kitchen brigade. The chart in this chapter's Math Screen shows the salary rates for many of the food service positions discussed in this chapter.

RELATED CAREERS

There are a number of careers related to food service. These careers include food service marketing, dietetics, sanitary engineering, health inspection, product merchandising, and a variety of others that you may wish to find out about from your school counselor.

Food Service Marketing

A **food service marketer** is involved in the creation, purchase, and distribution of products for the food service industry. Marketers participate in a wide range of sales and management activities, from the wholesale produce market to the trade show floor. A **purveyor,** or supplier, for example, identifies a customer's needs, develops sales strategies, and distributes products. Any item used by a food service operator needs to be marketed and sold by a food service marketer.

ILLUSTRATION 2–6 Food service marketers sell their goods at wholesale markets.

MATH SCREEN

On the accompanying graph, find a position that you have held or one in which you are interested. Use either the wage rate that you received or the one on the graph and calculate what your monthly and yearly gross pay would be if you worked a 40-hour week. The jobs of janitor, cafeteria server, and fast-food counter worker are often held by students as first-time jobs.

Note: Figures represent median hourly wages in dollars. For tipped employees, the figures given are cash wages excluding gratuities.

Source: Copyright *Nation's Restaurant News*

ILLUSTRATION 2–7 Food trade shows give marketers an opportunity to display new products and services.

Dietetics

A **dietitian** is a person with the nutritional background necessary to develop healthy and appealing menus for restaurants, institutions, or individuals with special needs. This demanding job requires a bachelor's degree, a passing score on the American Dietetics Association's registry exam, and a one-year internship. In a food service establishment, a dietitian is responsible for menu planning and development. To successfully manage an **institutional food service** such as a school or a hospital cafeteria, for example, contracted agencies will often have a restaurant manager and a dietitian work together to run the entire operation.

Sanitary Engineering and Health Inspection

Sanitation and safety techniques are necessary to keep a food service operation in compliance with federal, state, and local health and safety regulations. **Sanitary engineers** and **health inspectors** make periodic inspections for federal, state and local governments.

Product Merchandising

Creating consumer interest in food service outlets and products involves advertising in newspapers and magazines, public relations, and product development. Career opportunities in these areas include working as a food stylist, a food writer, or a product developer.

ILLUSTRATION 2–8 Sanitation and safety techniques are necessary to keep a food service operation in compliance with federal, state, and local health and safety regulations.

A **food stylist** designs plate and food presentations for advertising and public relations photographs and films. This person must have a high degree of culinary skills, be aware of the effects of time, heat, and exposure to different foods, and be up to date on current food trends.

A **food writer** must have good writing skills and culinary knowledge. For more on this position, see the career profile of food columnist Elaine Tait in Chapter 1.

A **product developer** needs cooking skills, creativity, and the ability to anticipate future food trends. Recent food product development innovations include many pre-portioned and pre-prepared items for microwave and convectional cooking. In addition, the need of food service businesses to control costs has created many product development job opportunities in the area of volume feeding.

EDUCATION FOR FOOD SERVICE PERSONNEL

Education for a food service career can be obtained in a number of ways. Most employees in the field receive their education through high school occupational home economics food service programs, vocational schools and college programs, or specialized training programs.

Occupational Food Service Programs

Home economics teachers offer food service and restaurant management classes at the high school level. These teachers are competent and knowledgeable; they are required to have at least a bachelor's degree and over two thousand hours of work experience in the food service industry. Most teachers also attend workshops and training programs to update their knowledge of restaurant procedures.

Food service teachers run active production kitchens and restaurants at their high schools for students interested in learning food service management. Students in these programs can learn food production and preparation skills as well as dining room service techniques.

ILLUSTRATION 2–9 Education for a food service career can be obtained in a number of ways.

Vocational Schools and Community College Programs

At the postsecondary level, food service teachers usually have graduated from a two-year culinary school program and must either have a bachelor's degree or be working on one. Many professional chefs pursue teaching careers at vocational schools and community colleges, going directly from commercial production kitchens into food service classrooms. Students in these programs thus have the opportunity to learn from experienced professionals. Vocational school and community college programs typically range between 18 to 24 months in length.

College and University Programs

College and university food service professors have master's degrees and, in some cases, doctoral degrees. In addition, they usually have extensive work experience in a specialty area such as restaurant, hotel, or institutional food service management. Most university programs now require students to complete a four-year bachelor's degree curriculum.

Training Programs

As the labor shortage intensifies and the need for qualified help increases, many food service organizations are initiating programs to train or retrain employees to perform new and improved food service techniques. Training specialists, working for equipment manufacturers or as independent consultants, design and conduct training programs for all levels of the industry. The qualifications for this position can be based on experience rather than education. The quality of training programs depends upon the communication, teaching, and organizational skills of the trainer.

FOOD SERVICE CAREER SUITABILITY INVENTORY

Are you suited to any of the food service positions discussed in this chapter? To find out, respond to the following questions. Place a check in the blank next to each question to which you would answer yes.

_____ Do you enjoy working with people while under pressure?

_____ Do you like serving people and treating them as guests?

_____ Do you mind working for long periods of time in less than ideal conditions?

_____ Are you willing to go out of your way to make people feel satisfied?

_____ Do you like immediate response from people for the work that you do?

_____ Do you want your daily work to be creative and constantly changing?

_____ Do you enjoy being involved with food in either the dining room or the kitchen?

If you answered yes to all of the questions, then you would probably enjoy a career in food service. You can explore food service career opportunities further with your school counselor.

CAREER PROFILE
JOE GENTILE: OWNER, SEAFOOD SHANTY RESTAURANTS

Joe Gentile says that his first restaurant job "was warm, there was plenty of food, and it paid well. With those incentives, it didn't take long for me to see that this business had a future."

A resort in the Catskill Mountains of New York State provided further experience. While working there one summer, Mr. Gentile performed a broad range of food service tasks. By the time he finished school, he was already developing restaurant concepts for investors. In 1970, circumstances allowed Mr. Gentile to put his ideas to work for himself. He opened the doors to the first Seafood Shanty Restaurant in Langhorne, Pennsylvania, with only $8 left in his pocket. The converted 19-seat hot dog stand specialized in serving fresh seafood. Mr. Gentile combined the restaurant with a retail seafood shop, and the Seafood Shanty concept now serves over fifty thousand customers weekly.

Whether he is involved with product development, menu marketing ideas, or new computer software programs, Mr. Gentile is involved in all aspects of his restaurants. This personal interest is the key to his success.

SUMMARY

This chapter has introduced you to the wide range of job opportunities offered by the food service industry. As you look ahead and plan for your education and future, you might want to further explore some of these job opportunities.

If you enjoy working with the public, for example, you might want to look into some front-of-the-house positions offered by restaurants. If food preparation is something you enjoy, you could perhaps become a part of an organization's back-of-the-house kitchen brigade. Many related job and career choices also are available, including dietetics, food service marketing, and product development.

A number of sources provide the education and experience necessary to pursue a food service career. Depending upon your goals and resources, you could receive training through high school occupational home economics courses, two- and four-year colleges, or on-the-job programs.

Working in the food service industry is both exciting and challenging. Perhaps your interests and skills match one of the many fulfilling jobs or careers in this dynamic field.

QUESTIONS

1. How do your family's weekly eating patterns compare with the national trends discussed at the beginning of this chapter?
2. What impact does the food service industry have on the job market in the United States?
3. Outline and define the positions in the kitchen brigade. Which of these jobs sounds most creative? most difficult?
4. What are four food service careers not directly involved in food production and service? List and describe them.
5. What is the definition of *food service management?*
6. Compare front-of-the-house and back-of-the-house positions. What is the difference between them in terms of job responsibilities?
7. Turn to the restaurant organizational chart in Figure 2–1. Which job positions are front-of-the-house jobs and which are back-of-the-house jobs?
8. What job position in the food service industry matches your personal goals, interests, and skills? Explain why this is a good match.
9. What does the term *pastilliage decorations* mean? Who prepares these decorations?
10. Describe one of the food service educational programs discussed in this chapter. Explain why this program would be most suited to your career goals.

ACTIVITIES

Activity One Design a graph to chart your family's eating patterns for a one-week period. Make sure to include all outside-of-the-home food consumption, including fast food and take-out food.

Activity Two Visit a restaurant and create an organizational chart based on that business.

Activity Three Choose one position from the organizational chart. Find someone who has that job and interview them. Make sure to find out their educational background, career goals, and current job description.

Activity Four From among the jobs discussed in this chapter, select one that appeals to you. Interview someone who holds that job in your community to determine why the person selected the job and whether or not he or she enjoys it.

PART TWO

SERVER SKILLS

3

RESTAURANT STYLES OF OPERATION

VOCABULARY

commercial
profit
institutional
nonprofit
full-service restaurant
table service
wait staff
bus persons
deuces
theme or concept restaurants
banquet service
cafeteria service
family style service
fast-food service
sushi
food halls
sneeze guard
side stations
chafing dishes

OBJECTIVES

After studying this chapter you should be able to do the following:

- Outline the components of the six styles of food service operation
- Discuss the merits of cafeteria service
- Define fast-food service and discuss its impact on the eating habits of the American public
- Explain how take-out service can help to increase restaurant profits
- Discuss the reasons for combining self-service and table service in many food service operations

Food service operations fall into two major categories: commercial and institutional. **Commercial** restaurants are businesses. They exist to make a **profit,** or monetary gain, for owners and investors. **Institutional** food operations are part of larger, often publicly-owned, facilities such as hospitals, schools, retirement homes, or prisons. Many institutional food services must make some profit to maintain their facilities and operations.

More than 10,000 private clubs throughout the country represent another type of food service operation that is neither fully commercial nor institutional. Clubs with limited memberships are considered nonprofit institutions for tax purposes. **Nonprofit** businesses provide a service to the community. Any profit that is made must be used to maintain or improve the business.

In order to operate efficiently and effectively, commercial, institutional, and nonprofit operations must follow the same basic guidelines: provide quality food and service to their customers and follow proper management techniques to ensure a successful business.

ILLUSTRATION 3–1 More than 10,000 private clubs throughout the country represent another type of food service operation that is neither fully commercial nor institutional.

OPERATIONAL STYLES

Food service outlets can be classified according to six different styles. Restaurants are either full-service, banquet, cafeteria, family style, fast-food, or take-out operations, and many organizations combine a number of these styles in a single business.

Full-Service Restaurants

A **full-service restaurant** has a sit-down dining room. In addition, a carefully designed interior decor establishes a relaxing, romantic, stimulating, or other type of atmosphere, depending upon the theme, mood, or setting of the restaurant. The restaurant's formality will determine whether or not there are linens, flowers, or candles on each table.

Full-Service Restaurant Table Service

Full-service restaurants offer **table service;** that is, they provide **wait staff**—waiters and waitresses—to take menu orders and serve food to guests. **Bus persons** clean and set up the tables and often assist in serving. Some full-service restaurants also provide other customer services, such as cocktail lounges and entertainment.

Full-Service Restaurant Meal Services

The number of meal services that a restaurant offers depends upon the organization's location, management policy, and type of cuisine. Many restaurants can provide only two meal services per day, and some offer only a dinner service.

Restaurants that prepare both lunch and dinner may take advantage of the typically slow hours between 4:00 p.m. and 6:00 p.m. by offering a discounted early dinner time to attract an area's older population. Other restaurants attempt to draw a younger crowd by serving weekend brunch on Saturdays and Sundays between 10:00 a.m. and 2:00 p.m.

Full-Service Restaurant Seating

Table seating in most full-service restaurants is at round, square, or rectangular tables. Restaurants offer **deuces**—two seats—or four, six, eight, ten, and twelve seats, depending upon their size and available equipment. The primary objective, of course, is to seat a large number of people in as many different seating arrangements as possible. Servers and guests, however, must have enough room to move around the dining room and to eat comfortably. Restaurant design and floor plans are discussed more fully in Chapter 9.

Full-Service Restaurant Cuisine

Full-service restaurants offer many varied types of cuisine, often based on a theme or concept. Cuisine in modern American restaurants includes a full range of international foods, from Italian pasta to Chinese egg rolls. **Theme or concept restaurants** are centered around entertainment or decor. Menus in these restaurants typically include illustrations that reflect the restaurant's theme.

Banquet Service

Banquet service is used for groups of customers and requires a specialized form of seating arrangement and table service, called Russian Service. For banquet service, a restaurant, hotel, or banquet hall provides large private rooms that groups can rent for a social or business function. The food service for these functions is planned and prepared in advance and is served to the entire group at once, either at the tables or buffet-style. Banquet service can be a very profitable business. Functions are booked months in advance, which gives management time to develop menus and plan for the correct amounts of food and staff, thus avoiding costly waste.

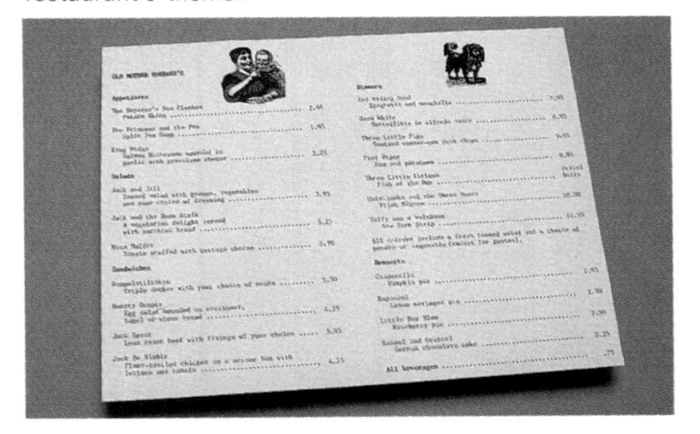

ILLUSTRATION 3–2 Theme or concept restaurants are centered around entertainment or decor. Menus in these restaurants typically include illustrations that reflect the restaurant's theme.

Cafeteria Service

Cafeteria service is usually associated with institutional and contract feeding (contracting an outside service to run a food service operation). Most Americans are first exposed to this type of self-service feeding in school. Cafeterias became popular in the United States during and after the Second World War, when there was a growing demand to feed large numbers of people quickly, efficiently, and economically. Hospitals, schools, the Armed Forces, manufacturing plants, and businesses in large office buildings are among the many organizations that continue to offer some type of cafeteria service on a nonprofit or low-cost, subsidized basis.

ILLUSTRATION 3–3 Businesses in large office buildings are among the many organizations that continue to offer some type of cafeteria service on a nonprofit or low-cost, subsidized basis.

Cafeteria Restaurants

For years, cafeterias have been a popular form of commercial restaurant in certain parts of the United States, particularly the South. The fast-food industry incorporated many aspects of the cafeteria concept, such as flexible table and chair arrangements to allow for as much seating as possible, during its rapid growth in the 1980s. Escalating operating costs, combined with la-

NUTRITION NOTE CAFETERIA SERVICE FOR OLDER ADULTS

During the 1990s, the percentage of the American population that will reach retirement age is expected to be larger than ever before. Many of these people will live in housing designed for retirees. Because cafeteria service is an effective, economical way to serve large groups of people, many of these facilities will incorporate that style of food service into their operational plans. The biggest challenge for these cafeterias will be how to feed a large group while still meeting the special nutritional needs of many individuals with various ailments or dietary restrictions. Most of these facilities will hire a registered dietitian to oversee menu planning and the development of special diets for those who need them.

bor shortages in the food service industry, will result in more creative adaptations of cafeteria and self-service feeding concepts.

Family Style Service

Family style service represents a very small percentage of the restaurant market today. Usually based upon a particular theme or cuisine, **family style service** allows diners to join others in a setting where large dishes of food are placed on a table and guests serve themselves. Health regulations in many communities restrict this type of table service because of possible food contamination and sanitation problems.

This format is also adapted for some forms of banquet service. Full-service restaurants sometimes adapt family style service for two, four, six, or eight guests. For this type of family style service, limited menus, including two or three entrées, two vegetables, a starch, salads or relishes, and a dessert, are common. Most Chinese restaurants would be classified as family style as well.

Fast-Food Service

Fast-food service utilizes efficient preparation and packaging techniques to produce foods for consumption at home or in the restaurant. A number of national chains have based their success on quick service and inexpensive sandwiches. The standard American hamburger is ideally suited to the fast-food concept.

Food Halls

The fast-food market has enlarged to include almost every type of food, from fruit freezes to Japanese **sushi,** a combination of raw fish and vegetables. Fast-food outlets have taken over a major portion of the retail space in shopping malls and downtown areas. Planners often group fast-food outlets together to form **food halls,** centralized areas for restaurants and seating. One advantage of this arrangement is the dynamic impact of the colors, displays, and smells produced by a variety of decors and foods. In addition, sanitation, cleanup, and waste removal efforts are confined to one general area. Food halls cater to a diverse cus-

ILLUSTRATION 3–4 The standard American hamburger is ideally suited to the fast-food concept.

ILLUSTRATION 3–5 Planners often group fast-food outlets together to form food halls, centralized areas for restaurants and seating.

tomer group by featuring a range of foods and beverages.

Take-Out Service

Take-out service can be a very profitable extension of any operation. To offer take-out service, managers do not have to add seating space, find extra help, create additional overhead costs, or buy more equipment. Take-out service provides the customer with prepared foods for at-home consumption. These foods range from appetizers and soups to entrées and desserts. Take-out service requires a well-planned packaging design that keeps hot food hot, cold food cold, and does not leak. A well-organized system for handling increased kitchen production as well as customer traffic in a designated pickup area is needed. Drive-thru service has become popular with many of the national fast-food operators.

FOOD SERVICE EQUIPMENT

Each restaurant style has different requirements regarding layout, seating, hours of operation, and service techniques. Table settings, for example, are much different in full-service and fast-food restaurants. The equipment used by most food service organizations, however, is common to many types of operations. The following sections describe the main pieces of equipment that support restaurant workers and ensure the safe and efficient distribution of food and beverages. This equipment includes self-service units, dessert dollies, service carts, side stations, and buffet service supplies.

Self-service Units

Many restaurants today combine both self-service and table service methods. In an effort to cut down on labor, create interesting dining themes, and reduce food costs, food service managers are incorporating self-service equipment into their dining rooms' traditional arrangements.

Salad Bars

One of the most common self-service units in restaurants today is the salad bar. Food service operations from cafeterias to fast-food and family style restaurants use a variety of salad bar designs. Salad bars provide a visual presentation of colorful foods kept chilled at a minimum temperature of 32–45 degrees Fahrenheit. Customers choose their own salad ingredients, select an appropriate dressing, and, in some restaurants, return as often as they want. The salad bar is also a form of entertainment, providing customers with a focus for conversation and a way to take an active role in dinner service.

Salad bars require constant attention to sanitation and visual presentation. In addition, these units create a high food cost if the individual salad ingredients go unused or are not placed properly in the bar.

State and local health regulations require that salad bars be equipped with a sneeze guard to prevent foods from being contaminated. A **sneeze guard** is usually placed at a level between a typical guest's chest and chin. This allows guests to reach food items, but not to breathe on them.

Hot-Food Bars

Hot-food bars have the same advantages as salad bars do in terms of saving on labor costs and increasing customer interest. Family style opera-

ILLUSTRATION 3–6 Family style operations often use hot-food bars to offer promotional all-you-can-eat buffets.

tions often use hot-food bars to offer promotional all-you-can-eat buffets. For these events, managers choose the food to be offered based on a balance of high-cost and low-cost items. One or two high-cost items are generally featured, surrounded by many low-cost items to help fill in the bar.

As with salad bars, the primary problems with hot-food bars are sanitation and visual presentation. Constant cleanup and close attention to food items are required to successfully present a hot-food bar. Portable hot-food bars have temperature controls and often infra-red lamps over each food well. Food is thus heated from both the top and the bottom. Hot-food bars must keep foods at temperatures above 140 degrees Fahrenheit at all times.

ILLUSTRATION 3–7 Dessert dollies, which are portable carts with several levels, allow dessert items to be displayed in a variety of ways.

Dessert Dollies

Dessert dollies, which are portable carts with several levels, allow dessert items to be displayed in a variety of ways. Whole cakes and pies, for example, are precut in the kitchen. Soft desserts, such as mousse and Bavarian cream, are presented in attractive serving dishes and glassware.

Rolling Carts

Table-side food preparation requires a rolling cart. The cart is an integral part of French table service and has been incorporated into American table service by many full-service restaurants. Rolling carts serve as portable side stations on which maîtres d'hôtel, captains, and servers prepare salads, entrées, and desserts.

Side Stations

Side stations are essential to any dining room, regardless of its operational style. **Side stations** provide a front-of-the-house work area where servers can store condiments, extra tableware,

ILLUSTRATION 3–8 Rolling carts serve as portable side stations on which maîtres d'hôtel, captains, and servers prepare salads, entrées, and desserts.

napkins, and cleaning equipment. They are also used as water and coffee stations when needed. Portable side stations are more versatile than those which are built into the room. They are movable for cleaning purposes and allow for a more flexible room arrangement.

Side Station Contents

Items included in a well-stocked side station fall into two categories. The first category includes a selection of glassware, flatware, ashtrays, menus, wine lists, linen or paper napkins, and serving utensils. The second category includes salt, pepper, sugar packets in holders, loose sugar in bowls, (if used), and a selection of all condiments available to guests. Additional equipment and supplies may be added, depending upon the dining room's menu and upon its style of operation.

Hot beverages are served from side stations so that servers do not have to go into the kitchen every time a guest requests a cup of coffee. Pitchers of ice water are also kept on the side station. Proper sanitation techniques are required to prevent food contamination. Open and loose food items, for example, should be served from a pantry area of the kitchen, never from the side station.

Bus pans are often stored on the bottom shelf of the side station. Servers and bus persons fill these pans with dirty dishes and then take them to the dishwashing area.

Buffet Service

Presentation and self-service are a primary part of buffet service. This service style requires the use of different types of equipment than that used for other service styles. Chafing dishes are used to keep hot foods hot for a buffet service. **Chafing dishes** use a source of liquid heat, such as Sterno. The four parts of a chafing dish are the stand, the water pan, the insert pan, and the cover. Food is kept warm by placing an insert pan over water heated by the liquid-heat source. This method keeps the bottom of the pan at an even temperature and eliminates contact with direct heat, which could burn the food. The cover helps to retain heat, and, if vented, releases moisture before it condenses.

CAREER PROFILE
GAIL FINAN: DIRECTOR OF ADMINISTRATIVE SERVICES, BRYN MAWR COLLEGE; 1989 RECIPIENT, IVY AWARD

Bryn Mawr

After more than ten years in her current position, Gail Finan still loves her job, and she wants others to share in her enthusiasm:

I would like to interest young people into considering a career in college dining services. It couldn't be more creative, exciting, and challenging. I have a 125-seat restaurant and a snack bar that serves fast-food items. We also serve 1200 students 20 meals a week, for 32 weeks out of the year, in a cafeteria setting. In addition, I run a summer conference resort with 1000 rooms, sports facilities, a fine-arts center, meeting rooms, and rolling green lawns within 20 minutes of the center of a major American city.

Ms. Finan began her food service career in the School of Hotel and Restaurant Administration at Cornell University. She moved into college food services at the University of Pennsylvania. In 1979 she joined Bryn Mawr College as Director of Food Services, and she has since developed their program into one of the finest in the country.

Active in the National Association of College and University Food Services, Ms. Finan has served as the organization's regional northeast president and is a national board member. In 1989, she was awarded the prestigious Ivy Award. The Ivy Award is given each year by members of the food service profession to one of their peers, in recognition of that person's achievement of excellence in food service.

SUMMARY

The variety of restaurant operational styles in the United States today presents an exciting range of opportunities for food service managers and employees. Management must constantly be ready to incorporate new ideas and equipment into service. They must constantly evaluate the different merits of the six styles of food service operation, determining whether a full-service, banquet, cafeteria, family style, or take-out operation—or some new combination of these styles—is best for a particular food service facility or situation.

The wait staff needs to be continually involved in menu changes and new equipment designs, such as innovations in self-service units or side stations. This enables them to respond effectively and efficiently to the kitchen staff as well as to provide guests with an enjoyable dining experience, however formal or informal. The wait staff's knowledge of the proper use and maintenance of equipment is vital to the success of any restaurant.

QUESTIONS

1. What are the differences between commercial, institutional, and non-profit food service operations?
2. Identify the type of food service operation that presents challenging opportunities in management and is neither commercial nor institutional.
3. List the six styles of restaurant operation. How are they different?
4. Banquet services are usually provided in what types of facilities and rooms?
5. What major problem does family style service present?
6. Why have fast-food restaurants become such a popular source of food?
7. Cafeterias serve what major function better than other types of food service operations?
8. Why is take-out service a way of increasing profits with minimal risk and expense?
9. What new concept in fast-food restaurant development is being incorporated into shopping malls?
10. What are the requirements for successful take-out service?
11. Why do restaurants use salad bars and hot-food stations to help serve their customers?

ACTIVITIES

Activity One As a class activity, create a theme for a buffet service. Outline the menu and plan how to purchase, cook, and present the food that you will serve. Collect as much of the needed equipment as possible, along with appropriate decorations, and set up a buffet display. With the help of your teacher, you may want to incorporate this activity with actual food preparation exercises.

Activity Two Visit some food service operations that offer take-out foods and gather a variety of containers used for take-out service. Discuss the ways in which they meet the criteria discussed in this chapter. If you have a microwave oven available, try, with the help of your teacher, to heat some appropriate food in the different containers. Do the containers keep food warm? Do they leak?

Activity Three Arrange a field trip to a local hotel. Ask the food and beverage manager for a tour of all of the facilities, including the kitchen. Look closely at all of the equipment they have, and then write a report on the different types of facilities that the hotel offers, especially focusing on their banquet room equipment.

4 THE TABLETOP

VOCABULARY

tabletop
place setting
linens
condiment
trenchers
earthenware
porringers
tankards
show plate
serviceware
flatware
escargot
porcelain
serviettes

OBJECTIVES

After studying this chapter you should be able to do the following:

- Outline the development of eating utensils and serviceware
- Describe the differences between the three types of American service place settings
- Demonstrate how to set a full-service French place setting, a banquet service setting, and a family style dinner setting
- Define the terms *flatware, glassware,* and *china*
- Select appropriate place settings for different styles of food service operations

The setting for a meal goes beyond the interior decor and atmosphere of a restaurant. The table decorations and place settings that greet a customer reinforce the general theme of a restaurant and its menu. They help to establish different degrees of formality as well as a feeling of elegance, diversion, or casual fun.

Tabletop is the term that refers to the implements and decorations presented on a dining table. This includes the **place setting,** which is composed of china, glassware, and utensils that have been selected according to operational style. **Linens** are the fabrics used on the tabletop. Linen colors, patterns, and styles are a major part of a restaurant's theme. Various table decorations, from flowers to condiment holders, round out the tabletop. A **condiment,** such as catsup, mustard, or relishes, is used to make food more savory.

This chapter first offers a brief history of the tabletop. Four main styles of contemporary place settings, plus their variations, are then described. These four styles include tabletops for American, French, banquet, and family style service. The chapter ends with discussions of serviceware and linen selection.

A HISTORY OF THE TABLETOP

As centuries have passed, each culture has developed its own social rules for conducting meals. People have progressed from eating with their fingers to using plates and flatware.

Early Tableware

By the Middle Ages in Europe, thick pieces of round, dried bread, called **trenchers,** were used as plates. They absorbed gravies and sauces, and diners often ate them at the end of the meal. During the fifteenth century, a square piece of wood with a hollowed out circle in the middle replaced the dried bread. The term trencher, however, was still used to describe this utensil.

Knives and spoons were unnecessary, as most meats and poultry were torn apart by hand and then eaten. Guests used their fingers to serve themselves from large bowls and platters. Good table manners required that guests wash their hands in public before a meal.

Catherine de Médicis of Italy introduced the fork to northern Europe when she married Henry II of France in 1533. This utensil, however, did not become an established part of the table setting until the late seventeenth century. Knives, forks, and spoons were considered personal items, and members of the upper class carried these utensils with them. The host or hostess was not expected to provide serving utensils for guests.

The Development of Modern Tableware

The first cups and plates were made of pewter or **earthenware,** commonly called pottery. In sixteenth-century England and France, glassware and silver appeared on the tables of the wealthy nobility. By the eighteenth century, fine-porcelain factories in France produced sets of dinnerware and serving pieces.

Nineteenth-century Europe provided the pattern for today's table settings. Serving utensils were designed for all types of food items. Linens,

ILLUSTRATION 4–1 The first cups and plates were made of pewter or earthenware, commonly called pottery.

laces, and ornate silver and gold table decorations also appeared during this century.

The American Table

The development of the American tabletop followed the European pattern. Early Colonial settlers brought a few treasured pewter **porringers,** or small bowls, and large mugs, called **tankards,** from England. Carved wooden bowls and plates, along with simple clay pieces, completed the typical tableware inventory. There were, however, excellent American silversmiths, such as Paul Revere, making decorative serving pieces.

PLACE SETTINGS

American restaurateurs often discuss place settings for different situations. Place settings are highly individualized in American restaurants. There are no absolute rules about how place settings should be set up for each service style, just as there are few firm rules about the execution of the different service styles themselves.

The following examples illustrate a wide variety of acceptable place settings. The figures show place settings that are standard for the food service industry and that offer guidelines for both management and servers. These place settings can be adapted to different situations to add a creative, decorative element to a restaurant's decor.

American Service Place Settings

American service offers a range of place settings depending on both the meal and the restaurant's operational style. Figure 4–1 shows the simplest

FIGURE 4–1 Basic American Service Place Setting

table setup. Used for breakfast or lunch, this setting involves only a dinner fork, knife, teaspoon, and water glass. The knife blade, as in all place settings, is placed toward the plate. The entrée is prepared and placed on a plate in the kitchen according to American service style and then brought to the table.

Figure 4–2 illustrates an all-purpose lunch and dinner place setting. As with the place setting in Figure 4–1, the main course is served on an entrée plate readied in the kitchen. A soupspoon is added to the utensils, and the knife, soupspoon, and teaspoon are arranged in the order that courses are served. This place setting also includes a salad fork, which is set on the outside, next to the dinner fork. A napkin is placed on the table.

Figure 4–3 shows the American full-service table setup for breakfast and lunch. China pieces include a bread and butter plate, a cup and saucer, an entrée plate, and a side plate for salad.

Figure 4–4 illustrates the American full-service dinner place setting, which is similar to the banquet service place setting shown in Figure 4–7. Note that the salad fork is placed to the left of the dinner fork in Figure 4–4, and to the right of the dinner fork in Figure 4–7. Either of these positions is acceptable.

FIGURE 4–2 All-Purpose American Service Place Setting

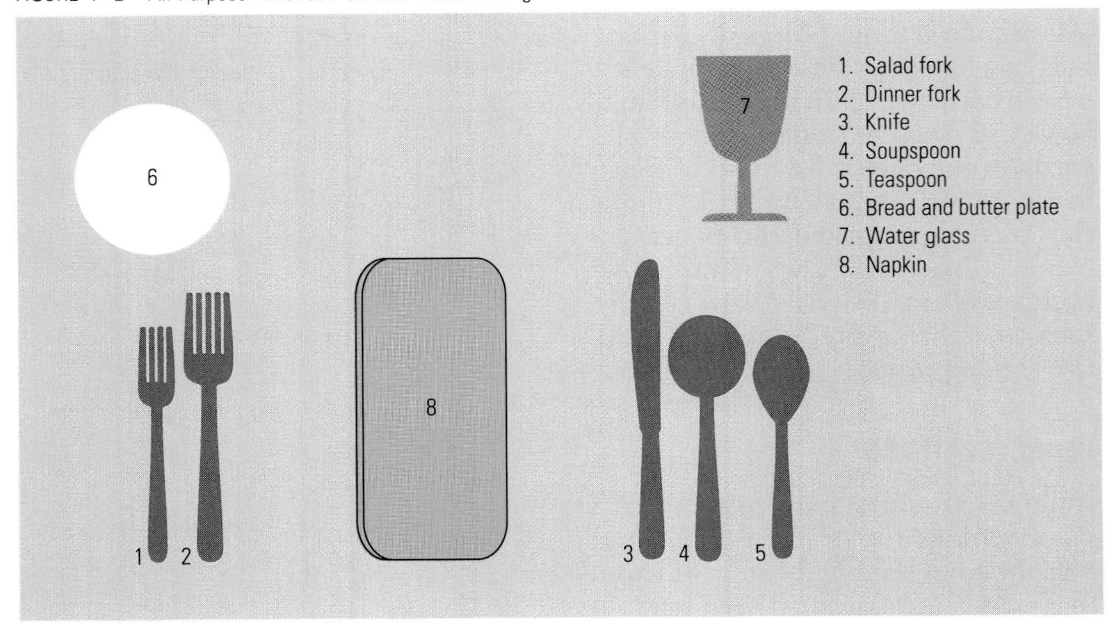

FIGURE 4–3 American Full-Service Place Setting for Breakfast and Lunch

FIGURE 4–4 American Full-Service Dinner Place Setting

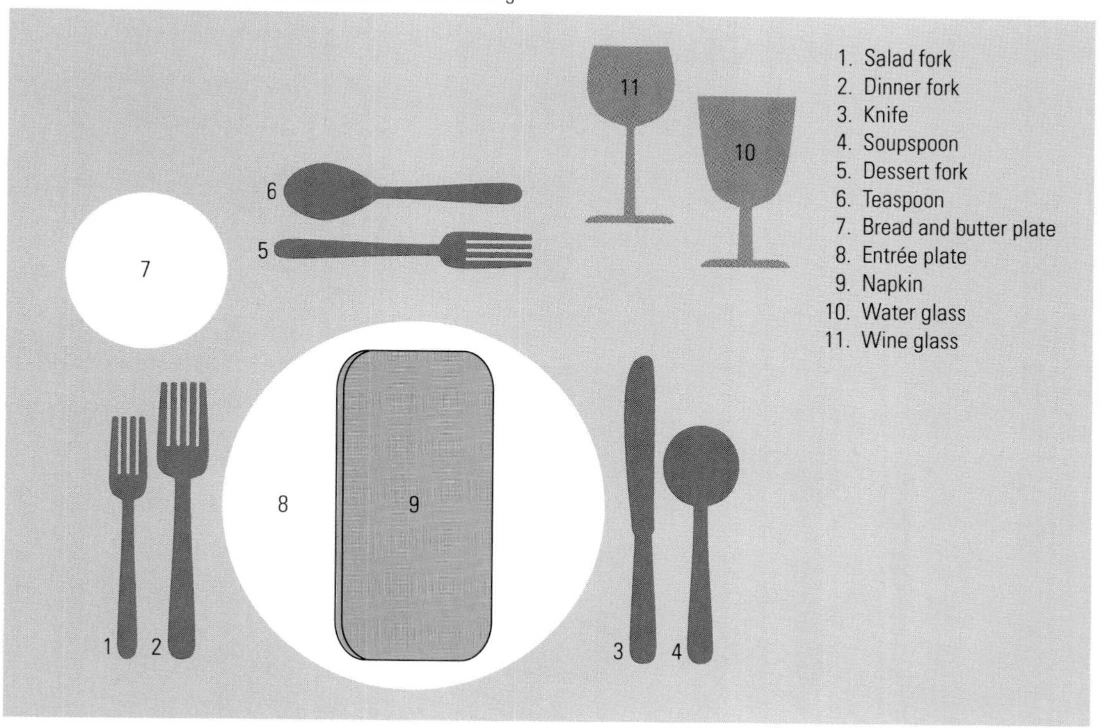

French Service Place Settings

In classical French service, flatware is placed on the table with each course, according to the food items that a guest orders. Wine glass selection is based on the type of beverage requested. In Figure 4–5 a white wine glass is preset on the table.

A French full-service place setting is shown in Figure 4–6. A white wine glass is the only glassware preset on the table. Most appetizers, salads, and soups complement white wine rather than red wine. If a guest changes to red wine during the course of a meal, the wine glass is replaced.

Banquet Service Place Settings

A banquet table setup generally consists of required flatware, a bread and butter plate, a napkin, a water glass, and wine glasses as needed. A cup and saucer is not usually preset in a banquet service place setting. The plate shown in Figure 4–7 is a **show plate**—a plate used for display purposes only. The actual dinner plate is set down following the removal of the appetizer course and just before the entrée is served. Note also that only the soupspoon is placed to the right of the knife. The teaspoon and dessert fork are positioned at the top of the

FIGURE 4–5 À la Carte French Service Place Setting

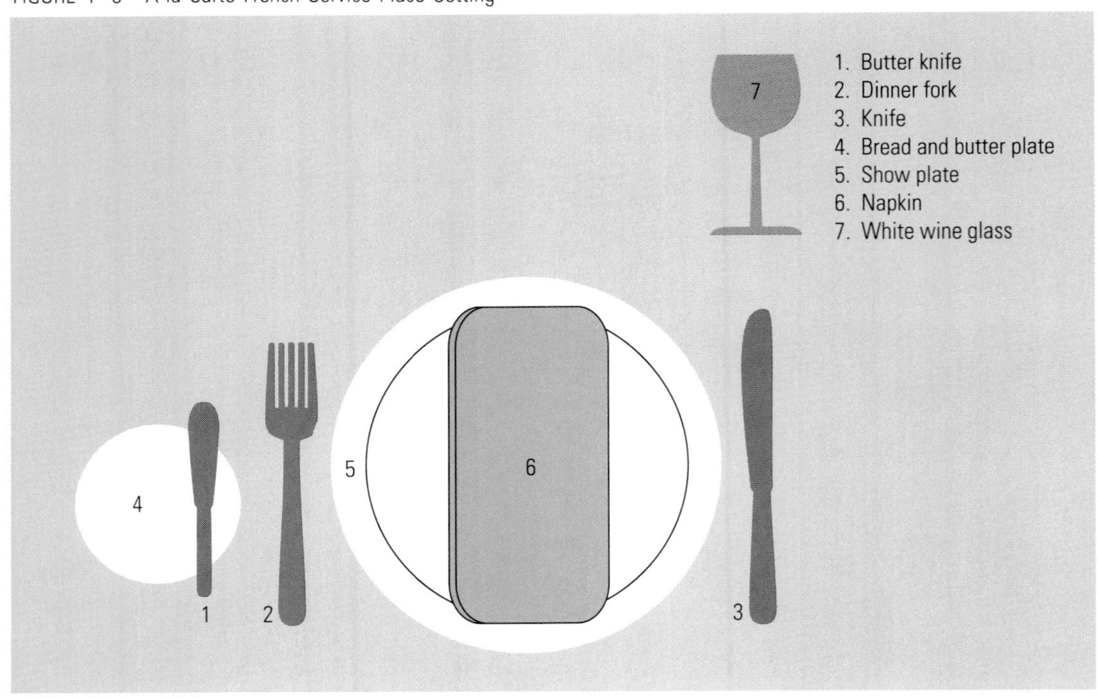

FIGURE 4–6 French Full-Service Place Setting

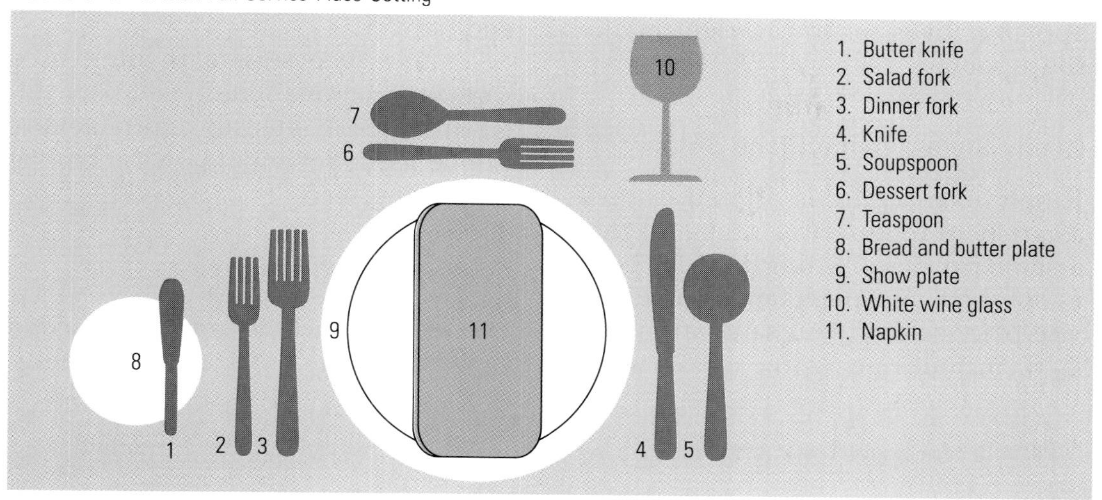

FIGURE 4–7 Banquet Service Place Setting

plate. If a teaspoon is needed for an appetizer, it is set to the right of the soupspoon.

Family Style Service Place Settings

Family style service is offered in such a variety of food service situations that a standard place setting has not been established. A typical family style service place setting, illustrated in Figure 4–8, includes an entrée plate, water glass, dinner fork, knife, teaspoon, and napkin.

Family style service is often used with international cuisines. Many different types of utensils and tableware can be used, depending on the type of food being served.

SERVICEWARE SELECTION

Serviceware includes any item used in the front of the house to serve guests.

FIGURE 4–8 Family Style Dinner Place Setting

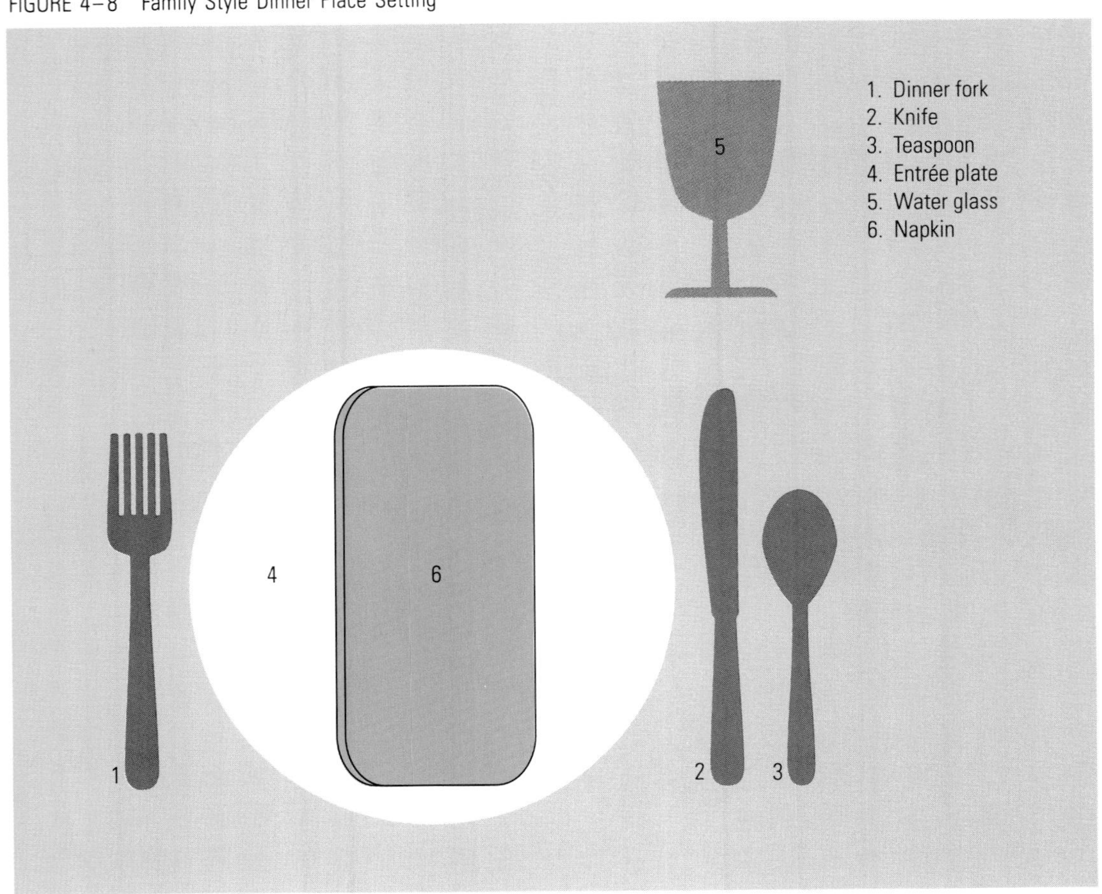

The three major categories of tabletop serviceware are flatware, china, and glassware. The following sections will discuss the general characteristics to consider when choosing different types of serviceware. Food service managers should evaluate serviceware pieces on the basis of durability, design, and sanitary features.

Flatware

The utensils used to serve and eat food are called **flatware.** Most food service outlets use flatware made out of a variety of metals. Fast-food and take-out operations use disposable plasticware.

The quality of a restaurant's flatware is directly related to the formality of its operation. Full-service restaurants, for example, usually use silver plate, while other, less formal food service businesses use anything from stainless steel to nickel-plated metal. Regardless of the material from which it is made, flatware pieces are chosen to suit a restaurant's operational style.

Flatware comes in a wide variety of shapes and sizes, some more familiar than others. Specialty flatware pieces are used only if certain items are on the menu. Particular utensils are necessary, for example, to eat foods such as shrimp and **escargot,** or snails.

ILLUSTRATION 4–2 Flatware comes in a wide variety of shapes and sizes.

China

Food service china is generally made from porcelain. **Porcelain** is the finest of the four bases that include terra cotta, stoneware, and pottery. Porcelain is made from a base of clay, feldspar, quartz, and, often, bone. To finish a piece of porcelain china, it is fired in a kiln at high temperatures. The surface of porcelain is nonporous and highly glazed. Intricate patterns in a wide variety of color combinations are painted onto most porcelain surfaces before they are glazed. Gold and silver are popular colors for restaurant china trims. Often a restaurant's logo or pattern is incorporated in their china. As with flatware, the design and durability of a restaurant's china depends on its formality and style of operation.

Glassware

Glassware for restaurant use can range in quality from fine lead crystal to inexpensive glass. Glass is made from a base of melted sand and is either blown or molded into shape. Some glass pieces are sturdier than others, and some manufacturers claim that their glassware is unbreakable or chip resistant. The glassware offered by a restaurant is relative to the style of food service operation. Casual dining restaurants generally use durable, utilitarian glassware. Full-service restaurants use more formal stemmed goblets for water and wine and a variety of glasses for beverage service.

ILLUSTRATION 4–3 Porcelain is made from a base of clay, feldspar, quartz, and, often, bone. Its surface is nonporous and highly glazed.

LINEN SELECTION

The standards used for choosing commercial linens are much different from those used to select linens for home entertainment. Restaurant fabrics must withstand, without shrinking or fading, the harsh chemicals and high temperatures used for commercial laundering. Food service managers thus choose linens primarily on the basis of durability. Napkin fabrics must provide a heavy-duty finish that is stain resistant and stiff enough to maintain a variety of folds. Linens can be commercially pressed to present a stiffer finish and hold together longer, but this process increases laundry costs.

Napkin Fabrics

The wide variety of napkin fabrics has ranged from the edges of rough-cloth table coverings to fine silk **serviettes** (the French term for napkins).. Most napkins used today are disposable. These napkins, often made of high-quality paper and available in attractive color combinations, are both easy to use and sanitary. Many restaurant and food service operations, however, continue to include cloth napkins as part of their overall tabletop design. Linen and other blended fabrics add a touch of elegance and sophistication to a dining experience.

ILLUSTRATION 4–4 The wide variety of napkin fabrics has ranged from the edges of rough-cloth table coverings to fine silk serviettes.

Napkin Folding

The art of folding napkins into attractive and decorative shapes is centuries old. Over time, different folds have become established and now have recognized names, such as the candle, the bird of paradise, the pyramid, the rosebud, the crown, the arum lily, and Lady Windermere's fan. Folded napkins are often used as table decorations on the entrée or show plate at each place setting. Fresh flowers or table gifts are sometimes worked into the presentation. Napkins can also be folded and placed in wine glasses and water goblets.

CAREER PROFILE

M. KATHIE DALRYMPLE: INTERIOR DESIGNER, J.K.D. & ASSOCIATES, WASHINGTON, D.C.

M. K. Dalrymple is an interior designer in Washington, D.C. Her clients include food service managers at racetracks, sports stadiums, restaurants, and sports bars.

With a bachelor's degree in interior design from California State University in Fresno and San Luis Obispo, Ms. Dalrymple started her interior design career in the contract department of a food service equipment company. With this company, she was involved with facilities planning for a variety of food service outlets, from fast-food to full-service restaurants.

Her next stop was the facilities planning and corporate artwork department of the R.J. Reynolds Company. From there she founded her own interior design firm, J.K.D. & Associates. One of her special skills is designing tabletops for restaurant clients.

In addition to her design work, Ms. Dalrymple is a licensed member of the American Society of Interior Designers (ASID) and an Institute of Business Designers (IBD) associate. She also serves on the District of Columbia Building Codes Advisory Board. This board influences restaurant design by reviewing concerns ranging from structural to occupancy codes.

SUMMARY

The importance of the tabletop is often overlooked in the day-to-day operation of a restaurant. A flexible tabletop design allows for different linen colors and table accessories that can give an operation a constantly fresh look. Designing a creative, attractive, and appropriate tabletop, based on one of the settings discussed in this chapter, should be done early in the planning stages of a new restaurant.

Choosing durable and easy-to-care-for flatware, china, and glassware is also essential. These items not only enhance a restaurant's atmosphere and operational style but must also stand up to daily wear and tear. The careful selection of suitable place settings and serviceware is thus important to ensure both the pleasure of a restaurant's guests and the smooth functioning of its dining room.

QUESTIONS

1. How do table decorations and place settings influence a customer's reaction to a restaurant?
2. What did people use as serving and eating utensils before the fork and spoon were introduced?
3. What influence did Catherine de Médicis have on modern table service?
4. What are the three main categories of serviceware and the three characteristics that must be considered before purchasing this equipment?
5. What are the three types of American service place settings? Draw and label sketchs of each of these place settings.
6. Of the place settings presented in this chapter, which three are the most complicated?
7. What are the major differences between the basic American service place setting and the French à la carte service place setting?
8. What purpose should napkin folds serve as part of a place setting?
9. When choosing napkins for restaurant use, what qualities should be examined?
10. How did eating utensils and serviceware develop throughout history?

ACTIVITIES

Activity One Reproduce the place settings presented in this chapter. Identify which style of food service operation would be best suited to each place setting.

Activity Two Visit some local restaurants and record the different types of place settings that they use. How many of the restaurants use decorative folded napkins? Which napkin folds do they use?

Activity Three With help from your teacher, hold a tabletop design contest. Divide the class into teams and develop restaurant themes for which to design the tabletops. Include table decorations as well as china, glassware, flatware, and linens. Have a restaurateur from your community judge the designs.

5
SERVICE STYLES

VOCABULARY

maîtres d'hôtel
table service
French service
brigade de service
commis de range
aperitif
appetizer
sommelier
commis debarrasseur
geridon
rechaud
Russian service
entrées
escoffier dishes
American service

OBJECTIVES

After studying this chapter you should be able to do the following:

- Discuss the development of the profession of service through the twentieth century
- Compare and contrast the five major styles of service used in food service operations today
- Outline the sequence of American service
- Discuss the problems that can result from table-side presentation
- Outline the standard rules for table service

By the late 1700s, Beauvillers, the most famous French restaurateur of the period, had helped to establish the restaurant as an institution in Paris. The success of his restaurant was based on the same four components used to maintain a first-class restaurant today. These components were a well-trained wait staff, an elegant dining room, a fine wine cellar, and a kitchen staff with the ability to prepare superior foods.

This chapter outlines the qualities needed by well-trained servers and their support staff. A skilled wait staff must be familiar with a variety of serving styles. They must be especially proficient in the service style used by the food service operation where they work. Regardless of operational style, a wait staff's capabilities are a key element to restaurant success.

THE HISTORY OF SERVING

The art of serving or waiting on tables began early in human history, the first time that one person was required to attend to another person. The skills associated with serving food and beverages are relative to the sophistication of those being served and the quality of the cuisine being offered. Basic serving skills, however, have not changed since a wait staff served the Pharaohs of Egypt. These skills have been embellished as food service has evolved in various international cultures, particularly in Western civilizations.

Until the late seventeenth century, table service of all skill levels was practiced in private residences by a staff that was either bonded (meaning owned), apprenticed, or hired. Most house staffs were simply paid with room and board. If wages were added, they were generally very menial. A house staff was always expected to be available, regardless of the time of day or the circumstances involved.

French Service History

Following the French Revolution in the late 1700s, many of the chefs and **maîtres d'hôtel,** or dining room managers, who were employed by the French nobility left France for other parts of Europe. These people took with them the traditions of table service begun by Catherine de Médicis of Italy when she married Henry II of France in 1533. As with their cuisine, the French prided themselves on the quality of their table service, which enhanced the pleasure of a meal. This pattern of service skills exists today under the term *French service.*

American Service History

The French had a major influence on the development of table service in America. The influx of the French Royalists to America following the French Revolution was one factor in this influence. Thomas Jefferson's travels to France, his interest in French foods and wines, and the widespread adaptation of the French style of dining were other factors. The influence of French service greatly improved the rudimentary and often rustic public and private dining methods practiced in American hotels and homes of that day.

Due to the democratic nature of American society, the profession and traditions of serving are not highly regarded. Consequently, European immigrants have generally established themselves in major cities as the maîtres d'hôtel of fine restaurants and hotel dining rooms. They have gathered around them other Europeans who respect the professional and financial rewards associated with table service.

STYLES OF TABLE SERVICE

Table service is the placement of food on a table and the style in which it is done. It involves setting up the table, serving food and beverages, clearing the table, and dealing with customers.

The five major styles of restaurant table service used in America today are French, Russian, American, buffet, and family style. The following sections discuss the steps involved in each of these serving styles.

French Service

French service is the service style around which most of the other table styles have evolved. It is the most detailed service style. In its purest form, it can involve up to six members of the dining room staff, who form a ***brigade de service,*** or "dining room brigade." The *brigade de service* for front-of-the-house staff is illustrated in Figure 5–1; Figure 2–2 in Chapter 2 outlines the *brigade de cuisine.*

Seating

In the sequence of French service, the maître d'hôtel selects a station, or section of the dining room, and assigns guests to a table. Depending on the restaurant, this seating choice can be significant in identifying the importance of certain guests either to the restaurant management or to other clientele.

Menu Presentation

The principal server, or ***commis de range,*** presents the menu, explains any additions, and takes any predinner beverage order. Predinner bever-

FIGURE 5–1 Dining Room Brigade *(Brigade de Service)*

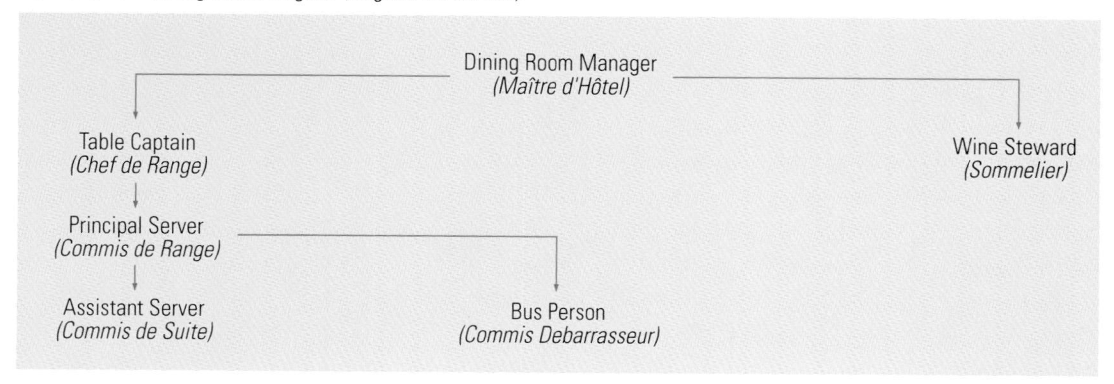

ages may be either a cocktail or an aperitif. An **aperitif** is an alcoholic drink ordered before a meal as an **appetizer,** a predinner food or drink that stimulates the appetite. Following the beverage service, the table captain will generally take the guests' food orders.

Wine Selection

After menu selections have been made, the wine steward, often called the ***sommelier,*** approaches with a wine list. The wine steward specializes in fine wines and knows which wines will complement different types of food, sauces, and spices. It is this person's job to help guests select wines that will enhance the flavor of each course. Many restaurants now serve wines by the glass rather than requiring the purchase of a full bottle. This practice helps both the wine steward and the guests to discover and enjoy a wide variety of wines.

Table Service

Classic French service includes an established course structure. Eight courses are served, beginning with an appetizer, which is followed by a soup and a fish course. An intermezzo sorbet is served as the fourth course. The fifth course is a meat. A salad, dessert, and cheese complete the meal. As each course is presented, the appropriate flatware is laid on the table (as discussed in Chapter 4). The *commis de range* and the ***commis debarrasseur*** (the bus person) are responsible for clearing each course and resetting the table for the next part of the service.

Table-Side Cooking

It is traditional in French service to finish cooking food in the dining room. Table-side cooking is used primarily for salad preparations, some fish and meat entrées, and flambéed desserts. In classic French service, however, table-side cooking is employed as much as possible, from the appetizer to the dessert course. The principal waiter works from a ***geridon,*** (a rolling cart), on which is placed a ***rechaud*** (a small open-flame heating unit), for table-side food preparation.

Russian Service

Russian service is generally used for banquets that involve table seating for 6 to 12 guests. All food is fully prepared in the kitchen and arranged on serving dishes. Servers bring food to the table and distribute it by portion to the guests. **Entrées,** or main course items, are placed on platters. Vegetables and starches are served in **escoffier dishes,** dishes with two compartments.

American Service

Predictably, **American service** is most prevalent in American restaurants. It is found in every type of operation, from family style to full-service restaurants. American service is popular because it dramatically reduces the number of servers needed in the dining room. All food is cooked and plated, or arranged on china, in the kitchen. The main course is placed on a dinner plate. The accompanying vegetables, starches, and garnishes are included either on the dinner plate or served in appropriate side dishes.

When American service is employed in a full-service restaurant or for a banquet service, servers use stainless steel plate covers to keep the food warm as it goes from the kitchen to the guests. One server, with a bus person as backup, can thus handle a number of dining room tables. American service can often minimize the amount of serving equipment needed, reducing the expenses involved with dish washing and upkeep. In addition, the amount of food served to each guest can be tightly controlled, which further helps make American service profitable by regulating food expenditures.

Buffet Service

Buffet service involves setting out a variety of food items on tables from

ILLUSTRATION 5–1 In American service, all food is cooked and plated in the kitchen.

which guests can help themselves. Most buffets are run on an all-you-can-eat basis, allowing guests to return as many times as they want. Some buffets offer carving stations and service staff to assist guests. Elaborate decorations are often added to increase visual appeal and reinforce a cuisine or theme. Small boats, wagons, and rolling carts can be used to display food items.

In informal situations, guests pick up flatware and napkins along with their plate. More formal buffets include place settings at each table and a wait staff to provide beverages and other services. In either case, the advantage of buffet service lies in significantly reduced labor expenses.

Family Style Service

Family style service is often called English service. In family style restaurants, all food is brought to the table on platters or in bowls. Guests pass the dishes and serve themselves with serving utensils. The sharing of utensils, however, greatly increases the possibility of spreading communicable diseases. Local health and sanitation laws thus prevent most restaurants from using this service style.

Modern Restaurant Service

In today's competitive restaurant industry, American restaurants seek to increase their efficiency and profits by modifying and combining the service styles discussed in the preceding sections. Few French restaurants in the United States, for example, employ a *brigade de cuisine.* Most service advertised as French service is actually a combination of French and Russian service, using table-side cooking and presentation as the highlight of a meal. Restaurants that primarily use American service often incorporate table-side preparation for salads, desserts, and beverages. Banquet service operations today often use American service techniques for side dishes, reserving Russian service for the main course alone.

Whatever a restaurant's style of service, it is important that all its servers be trained to present that style professionally. Nothing can ruin a restaurant's business more quickly than poor service. Servers thus must understand serving techniques and the sequence of service, and they must present food with knowledge and confidence.

SEQUENCE OF SERVICE

All full-service restaurants follow a general sequence in which different parts of a meal occur. This sequence depends on a restaurant's service style. Figure 5–2 lists the service steps in an American service restaurant.

The same sequence of service applies whether the restaurant's service style is formal or informal. Courses

FIGURE 5–2 Sequence in American Service

1. Welcoming the guest
2. Seating the guest
3. Presenting the menu
4. Taking the beverage order
5. Taking the menu order
6. Serving breads, butter, and relishes
7. Serving the appetizer course, which may be a soup
8. Clearing the appetizer course
9. Serving the salad course
10. Clearing the salad course
11. Serving the entrée and accompanying items
12. Clearing the entrée course
13. Presenting the dessert menu and selection
14. Serving postdinner beverages
15. Clearing the dessert course
16. Presenting the check

may be combined and service more rushed in an informal restaurant, but the meal is served in the same order, depending on which courses a guest requests.

RULES FOR TABLE SERVICE

Some generally accepted rules guide the way in which a restaurant's staff handles equipment, china, glassware, and flatware in the process of serving any type of meal. These rules fall into two categories: those that are standard for all restaurants, and those that are optional, depending on a restaurant's menu and operational style.

Standard Rules in Table Service

The following standard rules apply in every situation, regardless of how formal or informal the occasion or restaurant.

1. Food is served with the left hand from a position to the guest's left.
2. Beverages are served with the right hand from a position to the guest's right.
3. A table is served in the following order: women first, from oldest to youngest; children; and men.
4. When a dish is brought to the table, it should be set down in a position that will make it as attractive as possible to the guest. Depending upon the garnish, the appetizer and salad plates should be placed with the best possible visual presentation in mind. The main course should be placed so that the entrée item is directly in front of the guest.
5. Dishes are cleared with the right hand from a position to the guest's right.
6. When a guest moves a knife (blade facing in), fork, or both to the top of his or her plate, this action indicates that he or she is finished, and the plate should be cleared. Many guests, however, do not follow this tradition when completing their

ILLUSTRATION 5–2 The main course should be placed so that the entrée item is directly in front of the guest.

meal. Instead, the server and bus person may have to determine that the end of the course or meal has come by observing the amount of guest conversation and their general physical positions. If the server feels that the meal is finished, it is appropriate to approach the table and ask if plates may be cleared.

Optional Service Rules

Depending on the type of cuisine and service that a restaurant offers, there may be additional table service demands. The following situations require a number of service rules.

Salad Service

If a salad is being served as a separate course, it should be set directly in front of the guest. If it is being served with another course, such as the entrée, then it should be placed to the guest's left.

Dessert Service

To add to visual impact, servers may be asked to present desserts on a serving tray or a dessert cart. With tray presentation, the server can generally return to the kitchen to pick up a fresh, prepared portion of a requested dessert. Usually any dessert that is in a soft form, such as mousse or crème caramel, is already in an appropriate serving dish. When using a dessert cart, however, the server may have to take pieces of cakes, pies, or other desserts off the cart and place them on plates in front of the guests. All desserts on a dessert cart should be preportioned and sliced in the kitchen before being put on the cart.

ILLUSTRATION 5–3 To add to visual impact, servers may be asked to present desserts on a serving tray.

Table-Side Preparation

In addition to French restaurants, many full-service American restaurants offer various forms of table-side cooking. These can range from the assembly of a cold Caesar salad to the presentation of a flaming cherries jubilee dessert. Serving food with an open flame requires careful instruction and practice. Servers should never attempt any type of flamed item unless they have received sufficient training.

Servers also need proper instruction to prepare and present cold foods. Unless they are assembling an easy-to-prepare salad or dessert and adding a dressing or sauce, they will generally have to learn and practice a recipe several times before they are able to master it.

ILLUSTRATION 5–4 The most satisfying aspect of food service work is the gratification received from helping other people enjoy themselves.

THE SERVING PROFESSION

The most satisfying aspect of food service work is the gratification received from helping other people enjoy themselves. Creating pleasant experiences for others is basic to success in the food service industry.

The serving profession, however, can be demanding on an individual's time and energy. In commercial food service, it is not unusual to work six or seven days a week when business is active. The amount of time a server must spend at work depends upon seasonal demands and the current amount of business. If there is a labor shortage, whether chronic or occasional, servers sometimes have to work 12- to 14-hour days to complete their preparation and service tasks. It is often difficult to plan for personal needs because of the unpredictable nature of the business. In addition, food service workers know that when their friends are off work and having fun, they will often be working.

Despite these problems, the serving profession can be exciting, creative, fulfilling, and, often, financially rewarding. In the 1990s, more and more well-trained food service personnel will be needed. These professionals will command good salaries and benefits regardless of the type of operation at which they work.

CAREER PROFILE

JAMES RICE: DIRECTOR OF FOOD AND BEVERAGE OPERATIONS, EASTERN DIVISION, HILTON HOTELS CORPORATION

HILTON

Source: Hilton Hotels Corporation

Jim Rice has been a food service professional since he first joined the Canadian Pacific steamship line in the early 1950s. Born in Liverpool, England, he first went to sea as a bell boy on the *Empress of France.* Training at the Merchant Marine Catering School led to a position as *commis* in the dining room. Eventually Mr. Rice became a server in the first-class dining room. As a professional server on the *Queen Elizabeth 2,* he was encouraged to come to the United States.

At the age of 25, Mr. Rice arrived in New York City to begin work as a waiter at a Manhattan restaurant, Tavern on the Green in Central Park. He earned his associate's degree in hotel and restaurant management at New York City Technical College and joined Hilton Hotels Corporation in the early 1970s. In his present position, Mr. Rice oversees and directs food and beverage operations for all the Hilton Hotels in the eastern division of the United States. A sense of humor, professionalism, and, above all, constant attention to pleasing guests has led him to a successful career in food service.

SUMMARY

The serving profession has been a respected vocation for centuries in Europe. Serving began in the palaces and homes of European nobility and continues in the fine hotels and restaurants of modern European cities. Public dining began in the seventeenth century and grew as an entrepreneurial spirit flourished.

The American public has been slow to accept the serving profession as a desirable career. The last 30 years, however, have seen a significant change in this attitude due to the growth of the food service industry.

Servers are among the most important staff members in any food service operation. They have the most contact with guests and thus the greatest opportunity to provide a satisfying experience, however formal or informal the operation. Poorly trained servers can cause low sales and unhappy customers who never return.

Knowledgeable servers are familiar with the five major types of service—French, Russian, American, buffet, and family style—and proficient in the style used by the operation in which they work. They must also understand the sequence of service used by their restaurant, plus the standard rules of table service. For servers who meet these requirements, the opportunities for financial rewards and career growth are almost unlimited.

QUESTIONS

1. What four components have made for successful restaurants from Beauviller's time to the present?
2. What nation's people had a major influence on the development of American table service?
3. Which service style is prevalent in the majority of American restaurants today?
4. What is the definition for the term *table service?*
5. What are the five main service styles practiced in the United States today?
6. Outline the steps for the sequence of American service.
7. What are the differences between French and Russian service?
8. Outline the standard rules for table service.
9. What are the problems involved with table-side food preparation?
10. What are the definitions of the terms *maître d'hôtel, chef de range, commis de range, sommelier, commis de suite,* and *commis debarrasseur?*

ACTIVITIES

Activity One Research some of the different contributions that Thomas Jefferson made to American cuisine. Write a paper or design a presentation to explain them. You can use food history books and biographies of Jefferson as resources. Ask the research librarian at your school or local library for assistance.

Activity Two Identify a restaurant in your town or city that practices French service. How has it changed the traditional order of service?

Activity Three Choose a serving position and learn its responsibilities. Set up a mock restaurant or a mock serving situation in your classroom so that you and your classmates can try out different jobs in the serving profession.

6
CUSTOMER RELATIONS

VOCABULARY

courtesy
communication
suggestive selling
gratuities
diplomacy
Heimlich maneuver
TIPS
intoxicated
MADD
SADD
attitude
motivation
hygiene

OBJECTIVES

After studying this chapter you should be able to do the following:

- Identify the five major stages of meal service presentation that require well-developed communication skills
- Define the role and importance of communication in a server's job
- State the three major categories of customer safety, describe problems in each, and explain how to cope with these problems
- Discuss the importance of attitude to a server's success

Customers are the most important ingredient in a successful food service operation. Without customers to pay for food and service, a restaurant will fail. At one time, American consumers were willing to accept any quality of food or service in order to receive nourishment, but they are now much more discerning and demanding.

As American food service customers become even more demanding, the industry must respond to their needs. A growing awareness of the importance of customer service is having a significant impact on the success of food service operations. At the same time, the overall labor pool is shrinking, affecting the availability of qualified servers.

The management challenge of the 1990s is to train this limited number of servers to provide quality customer service. Management must recognize the vital role that servers play in the success of their operations. This chapter will discuss the skills a server needs to create good customer relations.

THE SERVER'S ROLE

The most important person in food service customer relations is the server. As the primary salesperson for the restaurant, the server communicates with the customer more than anyone else. The success of a dining experience is directly related to the server's ability to provide the quality of service that the customer expects. Customer expectations depend on the restaurant's operational style.

Management, unfortunately, often overlooks the key role that servers play in the success of their operations, and they thus fail to supply adequate training and support programs. The three skills that good servers need are the ability to be courteous, to communicate, and to provide appropriate table service. Serving skills are discussed in Chapter 5; the following sections concentrate on courtesy and communication.

ILLUSTRATION 6–1 The most important person in food service customer relations is the server.

Courtesy

The ability to display **courtesy,** or polished manners, is based on a respect for other people, regardless of their race, creed, national origin, or social status. Courtesy includes helpful or considerate acts or remarks.

Good manners are always the rule. Helping to seat guests, taking care of their coats, and anticipating their needs are basic parts of professional service. Recognizing older patrons, seating and serving women before men, and deferring to the host or hostess on decisions concerning the table are appropriate manners for a professional server. Terms such as *excuse me, please* and *thank you* are a server's basic vocabulary.

ILLUSTRATION 6–2 Good manners are always the rule. Helping to seat guests is a basic part of professional service.

Communication

Communication is the process of relaying and receiving information. A server's ability to communicate has a considerable affect on her or his success. A combination of a server's oral presentation skills, appearance, manners, and personality can influence a customer's overall perception of a restaurant. Good server communication skills result in greater guest satisfaction and an increased possibility that the guest will return.

The five components of meal service presentation that require good communication skills are introducing oneself, presenting the menu, taking the order, presenting the check, and solving problems. The following sections discuss these components and effective methods of communication.

Introducing Oneself

The first thing a server should do is approach the table and introduce himself or herself according to management policy. Generally the greeting is as follows, depending on the time of day: "Good evening, my name is (first name only), and I will be serving you tonight." Servers should immediately detect and react to any obvious guest needs. This could include seating for children, coat storage, table arrangement, and assistance of handicapped patrons.

Presenting the Menu

Communicating menu information to guests requires skills in public speaking, listening, and information processing. Several general rules of public speaking apply in this situation, including always speaking clearly, slowly, and distinctly. Servers should know and understand the words on the menu and pronounce them correctly. Speaking directly to guests and maintaining eye contact are also important. Servers must be effective listeners.

A server's ability to effectively communicate the menu directly controls a restaurant's sales. Restaurant customers are slow to question servers about menu items. They are frequently embarrassed to demonstrate a lack of knowledge. A server should be reassuring and anticipate needs by quickly and skillfully describing items and preparation methods. Higher average check amounts can be the result of servers who provide good menu information.

ILLUSTRATION 6–3 The ability to concentrate and to listen are essential skills when taking guests' orders.

Taking Orders

The ability to concentrate and to listen are essential skills when taking guests' orders. Servers must make sure to complete each guest's order before moving to the next guest. They should also repeat the order to confirm the guest's decision.

Servers are the primary salespersons in any restaurant. A good salesperson always looks for an extra sale. **Suggestive selling,** or mentioning items or features that a customer has not ordered, is a merchandising technique universally used in sales. In a restaurant, suggestive selling takes place when a server suggests menu items that a guest has overlooked, such as appetizers, desserts, and beverages. If ordered, these additional items can significantly increase the total check amount. Helping with menu selection is also part of suggestive selling. Managers often want their servers to promote high-profit menu items and daily specials. Being able to effectively sell these items can increase both the restaurant's and the server's profits; **gratuities,** or tips, are based upon the total check amount.

Presenting the Check

Modern check preparation is done by computer in many restaurants. In such restaurants, servers enter orders into a computer through terminal boards located in the dining areas. Guest checks are printed automatically at the close of the meal, leaving little room for server error in check totaling. (Food service computer systems are discussed in Chapter 21.) There are, however, some operations that still rely on handwritten checks. Examples of handwritten checks are included in this chapter's Math Screen. Each of these examples represent popular multipage carbon forms that allow adequate space for writing notations as well as guest orders.

ILLUSTRATION 6–4 Modern check preparation is done by computer in many restaurants.

```
----------------------------------------
        REMANCO HAS VISION!

         Remanco welcomes
            PIZZA HUT!

SERVER 1035 TABLE   601/ 1  TIME 13:57

      1 Chz Priazzo                6.99
          Small
      1 Lg Chz Pairs              14.99
      1 Diet Pepsi                 0.89
          Medium
      1 Iced Tea                   0.79
      1 Basket Stix                0.99
                               --------
                 NET SALES        24.65
               Beverage            0.11
                     TAX           1.38
                               --------
               GRAND TOTAL        26.14

 Thank you for your patronage.

 Visit our newest Pizza Hut on...

6/29/90 GUEST 2 NUMBER 11 REPRINT 1
----------------------------------------
```

MATH SCREEN

Calculate the correct totals for these sample handwritten checks. Be sure to pay close attention to the number of each item ordered. For sales taxes, add 6 percent of the check totals.

A.

CHECK NO.	DATE	TABLE NO.	SERVER	NO. PERSONS
1001	3/27	4	6	4

1		NACHOS GRANDES	3.95
2		POTATO SKINS	3.80
3		CHILI RELLENO	4.95
4		CHEF'S SALAD (BCD)	4.90
5		SEAFOOD SAMPLER	7.50
6		CHICKEN TARRAGON	5.50
7			
8	2	COKE @ .85	
9	1	TEA W. LMN @ .75	
10	2	COFFEE @ .75	
11			
12			

CHECK NO.	DATE	NO. PERSONS	AMOUNT OF CHECK
1001			

B.

CHECK NO.	DATE	TABLE NO.	SERVER	NO. PERSONS
15001	3/27	8	4	3

1		ZUCCHINI	2.25
2		BEEF FAJITAS	5.50
3		TUNA SALAD	4.95
4		DELUXE TURKEY PLATE	4.80
5			
6	2	ONION R. @ 1.75	
7	1	FRIES @ 1.25	
8			
9	2	STRAB. SHORTCAKE	
10		@ 2.50	
11	2	COLA @ .85	
12	3	MINERAL WATER @ 1.25	
13	3	COFFEE @ .75	
14		FOOD	
15		BEVERAGE	
16			
17		TAX	
18		TOTAL	

CHECK NO.	DATE	NO. PERSONS	AMOUNT OF CHECK
15001			

1. In check A, what is the total amount charged for food?
2. In check A, how many guests are being served?
3. In check B, what is the total beverage charge?
4. In check B, what is the sales tax?
5. In check B, what would be the gratuity, based on a 15 percent service charge?

Closing the service is the final communication between a server and guests. Servers should thank guests for their patronage and encourage them to return again soon. Managers may also want servers to offer guest comment cards for completion. In addition, servers should thank the host or hostess for their help.

Solving Problems

Handling guest problems requires **diplomacy,** skill in tactfully dealing with people, as well as courtesy. A professional server can often anticipate problems related to behavior, intoxication, or service, and they can thus take precautions to prevent such problems. Nevertheless, guests may become upset because the quality of the meal or service differed from their expectations. A range of other influences, over which the server has no control, may also cause a guest to become disruptive.

There are two cardinal rules to remember in handling food and service-related problems. These rules are *the customer is always right* and *never argue with a guest.*

If a problem is alcohol related, servers should call on management to deal with the guest. Servers who have taken the TIPS (Training for Intervention Procedures) program may be able to anticipate and solve many different alcohol-related problems themselves. (The TIPS program is explained in the following section.)

Diplomacy and courtesy are the two key communication skills involved with successful problem solving. Although it is often difficult to know how to handle many situations, servers with good communication skills can successfully handle many problems.

CUSTOMER SAFETY

Customer safety is a major part of customer relations. The way in which management and the service staff react to emergency situations can determine the seriousness of an accident or illness. Good restaurant managers always plan in advance for emergencies and establish standard procedures to deal with them. Servers are involved in three categories of customer safety. These categories are accident prevention and response, alcohol awareness, and illness treatment.

Accident Prevention and Response

Most accidents are unnecessary and due to carelessness. Falls are the most common type of accident. By thinking ahead and correcting problems as they occur, accidents can be prevented. Servers should be trained in first aid methods and fire safety. Good training programs teach servers how to use the equipment needed to extinguish all types of fires. (The three major classes of fires and the methods for extinguishing them are discussed fully in Chapter 8.)

Choking is a common form of accidental death. All servers should be trained in the **Heimlich maneuver.** In this first aid procedure, pressure is applied below the rib cage to expel food clogged in the esophagus.

Alcohol Awareness

During the 1980s, many groups brought increasing attention to the problems caused by drunk drivers, which led to questions about the legal responsibilities of organizations that serve alcohol. In response, the food service industry instituted alcohol-awareness programs. The major such training program for servers is called **TIPS** (Training for Intervention Procedures). TIPS helps servers identify when a guest is in danger of becoming **intoxicated,** or drunk. Alcohol affects the central nervous system and can impair a drinker's judgment and reflexes. The following quote from a TIPS brochure describes the benefits of the program:

The first step in dealing with the effects of intoxication is understanding what causes it. TIPS training helps participants understand how what's going on inside the body influences what's going on outside. Alcohol-serving professionals develop a "sixth-sense" about customers who are on the verge of becoming intoxicated. TIPS training shows them how to use intuition as a springboard to action, teaching strategies that can stop an overdrinking problem before it has a chance to start. TIPS demonstrates how the total environment of a bar or restaurant can help customers drink more responsibly. TIPS training can be especially valuable, teaching participants how to prevent a customer from becoming intoxicated with firmness, concern, and respect. No student in the program is certified without effectively participating in role playing to test their judgment and help them apply what they've studied. A written examination is also required for certification, which must be renewed every three years. TIPS training saves lives, keeps families intact, eases the financial burden on local communities, and helps keep insurance rates down. It's one solution to the problem of drunk driving.[1]

TIPS training also teaches servers effective techniques to prevent customers from driving under the influence of alcohol. These include talking customers out of driving, helping them find another form of transportation, and offering free food and coffee to slow down the effects of alcohol.

Drunk drivers cause thousands of deaths in the United States every year. Organizations such as Mothers Against Drunk Driving **(MADD)** and Students Against Drunk Driving **(SADD)** work through community organizations, advertising, and school programs to promote alcohol awareness and training programs such as TIPS.

Illness Treatment

The types of illnesses that occur most frequently in food service operations

[1] *Life Saving Training for Alcohol-Serving Professionals* (New York: Insurance Information Institute, April 1988).

are fainting, seizure, and heart attack. All these conditions are serious, and restaurant personnel should call for medical assistance. A heart attack victim, however, needs immediate first aid. Cardiopulmonary resuscitation (CPR) techniques should be performed while waiting for paramedics to arrive.

Whenever possible, it is advisable to get medical information from an ill guest before help arrives or the guest loses consciousness. Servers should also be aware of the legal liabilities involved with applying first aid. A server should never attempt a first aid technique that she or he has not been trained by professionals to apply.

Effectively handling accidents, alcohol-related problems, and guest illnesses demonstrates responsible food service management. Servers' technical skills should be augmented with safety awareness, TIPS, and first aid training.

THE SUCCESSFUL SERVER

Being a successful server takes more than the ability to communicate and a knowledge of table service techniques. Attitude, personal hygiene, and the ability to work as part of a team are just as important. How servers feel about themselves and their work is reflected in their appearance and in the way in which they work with those around them. A successful food service operation is dependent on the attitude of the people who work there. Customers react as adversely to a server's bad attitude as they do to poor quality food and service.

Attitude

Attitude refers to a person's emotional state and reflects the feelings that he or she has acquired about something or someone. Food service managers are primarily concerned with personal attitude and work attitude. The emotions that servers feel toward themselves and their work directly affect their productivity and the attitudes of those around them.

Personal Attitude

How people feel about themselves determines how happy they are both privately and on the job. Liking oneself is primary to having a healthy attitude. If a person feels comfortable with himself or herself, then others will also. A healthy attitude is reflected in the way that a person smiles and talks to people as well as in his or her personal appearance and posture.

Self-confidence is a large part of a successful person's attitude. If a person is happy with and confident in what he or she is doing on a job, then that person will be more tolerant of other employees and more patient in accepting the problems that others may impose on his or her work efforts. Being able to help other people do a better job and to work as part of a team are important personal as well as professional skills.

Work Attitude

In an industry as people oriented as food service, an employee's personal attitude becomes a vital part of his or her work attitude. Self-confidence is reflected in a worker's ability to sell and provide quality service to customers. An important part of developing a healthy attitude about oneself is being happy with one's work and goals.

A good attitude is directly responsible for job success. If an employee does not want to be on a job, then he or she will not perform to anyone's satisfaction, either the customer's or the employer's. One worker's bad attitude can affect other employees and their productivity.

Attitude problems can stem from personal as well as on-the-job factors. Illness, family problems, lack of communication between management and employees, and disputes with other workers can all affect a person's attitude. Employees who set personal problems aside while on the job and work harmoniously with other workers are happier with themselves and their work.

A good work attitude is seen in habits such as being punctual, following directions, responding to incentives, working as a team member, and doing the needed "extras" without being asked. Pride in a job well done is the foundation of a good work attitude. This pride comes across in a job interview as well as in performance, and it helps an employee to advance both financially and professionally.

Motivation

The need or desire that causes a person to do something is known as **motivation.** Pride is evident in highly motivated people. Motivated employees enjoy what they are doing and know where they are going both professionally and personally. Having ambitions and setting goals are important parts of a healthy personal and work attitude. The satisfaction of reaching goals and doing a good job leads to self-confidence, the key to developing a healthy attitude.

Uniforms

Uniforms establish a server's identity and reflect a food service operation's style. Uniforms create a dress standard, and they should be coordinated with a management policy on footwear and accessories. Appropriate uniforms are important because they create the first impression between customers and servers. The way in which a server's uniform and shoes are cleaned and maintained reflects the attitudes of both the staff and management.

Managers and corporate policymakers should choose uniforms not only for their design but for their fabric as well. The basic criteria for uniform selection include a suitable design, a comfortable fit, and an easy-to-maintain fabric. Cotton and polyester blends are the most popular fabrics for servers' uniforms.

Accessories should be selected with

ILLUSTRATION 6–5 Uniforms establish a server's identity and reflect a food service operation's style.

food sanitation in mind. Jewelry should be limited and, when worn, simple. Heavy costume bracelets and necklaces are inappropriate and increase the danger of food contamination.

Shoes should be chosen for comfort. In addition, the soles should be slip free to help avoid falls. Shoes with heavy-duty, reinforced toes can help prevent injury from dropped objects. Because servers spend so much time on their feet, they need good working shoes specifically designed for food service.

Personal Hygiene

Servers' personal hygiene standards are an important part of a food service operation's overall sanitation program. **Hygiene** is the science of health maintenance through cleanliness and the use of effective sanitary procedures. A server's personal hygiene and appearance give customers their first and most important impression of an operation's commitment to hygiene.

Good physical health in large part results from a positive attitude. When a person feels good, they want to look good. The basic components of good health are a well-balanced diet, exercise, an adequate amount of sleep, and regular dental and medical checkups. In addition, many people use a daily checklist to ensure that they establish and practice proper hygiene.

Good personal appearance is an important part of success in the serving profession. Looking good and feeling good go hand in hand to create a positive, self-confident attitude.

CAREER PROFILE

DONALD SCHOENBRUN: DIRECTOR OF TRAINING SYSTEMS DEVELOPMENT, POPEYE'S FAMOUS FRIED CHICKEN & BISCUITS, AND PAST INTERNATIONAL PRESIDENT, INTERNATIONAL FOOD SERVICE EXECUTIVES

INTERNATIONAL FOOD SERVICE EXECUTIVES ASSOCIATION

Donald Schoenbrun began his career in hotel and food service at the age of 14, when he began working, from the bottom up, at a summer resort hotel. From housekeeping to reservations, Mr. Schoenbrun held entry-level positions throughout the hotel, gaining job experience by working in every major area of the resort.

After being graduated from the University of Vermont with a bachelor's degree in hotel restaurant management, Mr. Schoenbrun joined the accounting firm of Laventhol and Horwath, based in Chicago, Illinois. He then moved to New Orleans, Louisiana, as a food and beverage cost controller. As general manager of the Prince Conti Hotel in the French Quarter of the city, he put his hotel background to good use. Recognizing the need for well-trained service staff in New Orleans' rapidly growing hotel and restaurant industry, Mr. Schoenbrun developed Louisiana's first hotel and restaurant management training program at a community vocational school.

In 1974 he joined Popeye's Famous Fried Chicken & Biscuits as director of training, creating programs, videos, and manuals for this major fast-food chain. In addition, Mr. Schoenbrun is involved in many of the hospitality industry's professional organizations. The American Society for Training and Development (ASTD) as well as the Council on Hotel and Restaurant Training (CHART) found him to be an active member. In 1985 Mr. Schoenbrun became the international president of the International Food Service Executives Association (IFSEA), the oldest professional organization for food service managers. Committed to the education and development of its members, IFSEA annually awards over $100,000 in college scholarships.

Mr. Schoenbrun's experience and education in both the food service and hotel management aspects of the hospitality industry give him an ideal background for hospitality-related training. His love of teaching as well as his concern with high-quality hotel and restaurant service have led him to a successful career dedicated to education.

SUMMARY

Good customer relations are essential to the success of any food service operation. The ability of a server to communicate is a key part of any restaurant's customer relations and sales efforts. A menu only lists the food and beverage items for sale. It is the server's role as salesperson to explain the menu, make foods seem interesting and exciting, and thus increase sales. Being courteous and diplomatic during all stages of a meal service are also important server skills.

A good server has a knowledge of safety rules and illness treatment as well as a concern for the customer's well-being. Alcohol-awareness training programs, which teach servers how to handle intoxicated customers, are also a part of responsible customer relations today.

Positive personal and work attitudes are also important ingredients in a server's success. The way in which a server treats restaurant guests reflects her or his self-image. Good personal habits that ensure a neat, clean, and attractive physical appearance help to create self-confidence and a good attitude. Servers play an important role in presenting management's attitude about both their operation and their customers.

QUESTIONS

1. What is the most important ingredient in a successful food service operation?
2. What is the major customer relations challenge for restaurant managers in the 1990s?
3. Why are servers the most important restaurant employees in terms of customer relations?
4. Besides serving techniques, what two major skills should servers concentrate on developing?
5. Why is communication such an important server skill?
6. What are the five components of meal service presentation?
7. What are the three main categories of customer safety?
8. What do *TIPS, MADD,* and *SADD* stand for?
9. Why is a good attitude so important to an effective server?
10. What are the two attitudes, as described in this chapter, that a server needs to develop to be successful?

ACTIVITIES

Activity One Find out if there is a SADD chapter in your school. What are the activities that this organization promotes? Develop an alcohol-awareness program that your class could present with SADD.

Activity Two Divide your class into teams. Act out different server-guest situations, based upon interactions discussed in this chapter, to test the "servers" for appropriate behavior when confronted with rude guests, emergency illnesses, and so on.

Activity Three Design a personal hygiene poster for employee training. Visit a local restaurant and evaluate the servers' sanitation practices, based upon information in this chapter and the section on food handling in Chapter 7. Present your findings and the poster to your class.

PART THREE

SANITATION AND SAFETY

7 SANITATION

VOCABULARY

sanitation
foodborne intoxication
foodborne infection
toxins
bacteria
Staphylococcus aureus
Clostridium perfringens
Salmonella
Clostridium botulinum
parasite
Trichinella spiralis
pesticides
preservatives
sodium nitrite
galvanized
FIFO

OBJECTIVES

After studying this chapter you should be able to do the following:

- Explain the role of each of the four federal agencies that are responsible for sanitation standards and regulations
- Identify the principle causes of foodborne illness
- Describe the factors associated with foodborne contamination
- Recognize the importance of insect and rodent control
- Analyze the significance of time and temperature guidelines as they apply to food sanitation

Food contaminated by bacteria, chemicals, or pesticides is a serious concern for food service operators. Restaurant managers must ensure that food served to customers is free from harmful organisms or toxic materials. The food service industry is responsible for protecting the public by training employees in the principles of sanitation. The responsibility for sanitation standards is shared by local, state, and national agencies. These agencies inspect, grade, and regulate food products in order to safeguard the public.

SANITATION

In food service establishments, sanitation means wholesome food handled in a clean, germ-free environment by healthy workers. **Sanitation** thus refers to the practices that keep food free from disease-causing microorganisms.

Each year 6.5 million Americans suffer from food poisoning, and of those stricken, 40 percent claim to have eaten contaminated food at a restaurant. The symptoms of foodborne illness resemble the gastrointestinal distress that is associated with the common flu. In fact, people often mistake food poisoning for a "24-hour flu." The total cost of foodborne illness in the United States is between $1 billion and $10 billion a year. This figure includes medical expenses and time lost from work. Contaminated food products, pesticide residues on fruits and vegetables, and eggs contaminated with *Salmonella* bacteria are major public health problems. These problems especially concern the people who produce, handle, and serve food to consumers.

Food Service Regulations

Recognizing the dangers posed by careless food handling, federal, state, and local governments have passed many regulations to ensure that food service establishments are in compliance with sanitation standards. These written regulations and standards are enforced through inspections and licensing. The government's main concern is to protect the food supply before it gets to the consumer.

From the time a food service establishment is planned, sanitation must be a major consideration. All features of a restaurant's physical plant should be constructed with cleanliness in mind. Installing adequate utilities, eliminating pest entryways and breeding places, and laying out equipment for easy cleaning are some important design considerations.

Inspections

Before a food service operation opens, a local health inspector visits the facility. An inspector approves or disapproves the opening of a restaurant, depending on whether the operation meets minimum sanitation standards. After an initial evaluation, visits can range from once a month to once a year. The frequency of subsequent visits depends on the workload of local agencies and the severity of violations detected on previous visits. A health inspector completes a report the same day as a visit, and the owner or manager of the facility retains a signed copy with a total score and a list of any violations. Management should study this report carefully and correct violations as soon as possible. Failure to correct violations may result in a fine, unfavorable publicity, or in extreme cases, closure of the establishment.

Four government agencies are involved in sanitation standards and regulations. The Food and Drug Administration (FDA) and the Centers for Disease Control (CDC) work with the U.S. Public Health Service (another federal agency concerned with citizens' overall health). The FDA is concerned with protecting the public from unsafe drugs and food additives, and the CDC provides assistance in identifying disease causes. The Environmental Protection Agency (EPA) and the National Sanitation Foundation (NSF) ensure public safety by protecting air and water quality and enforcing adherence to sanitation standards.

Food service managers should know and understand all federal, state, and local regulations. Copies of these regulations can be obtained from a local health department. Managers must make sure that all their food service personnel are aware of potential sanitation problems. This knowledge enables workers to carry out acceptable sanitary practices when purchasing, preparing, holding, transporting, serving, and storing food.

FOODBORNE ILLNESS

The two types of foodborne illnesses are **foodborne intoxication** and **foodborne infection.** Intoxications are caused by **toxins,** poisonous substances, formed in a food prior to con-

ILLUSTRATION 7–1 The Centers for Disease Control work with the U.S. Public Health Service, providing assistance in identifying disease causes.

sumption. Infections are caused by bacterial cells within an individual's gastrointestinal tract. Foodborne infections are the result of eating foods containing harmful bacteria or parasites. Ninety-nine percent of the microorganisms in foods are beneficial to the body—only one percent are dangerous. Disease-carrying microorganisms are transmitted to food or water by insects, animals, and humans.

The principal causes of food contamination are infected food handlers, contaminated food supplies, unsafe food handling, unsanitary equipment, and hazardous chemicals (see Figure 7–1 for a chart of sources). The ultimate goal of a sanitation program in any food service organization is to protect guests from foodborne illness. This is done by protecting food from contamination and reducing the effects of tainted food.

Bacterial Growth

Bacteria are microscopic plants living in many different forms. Some bacteria are harmful to humans, plants, and animals. Bacteria are so small that they cannot be seen by the naked eye. In fact, about ten million of these tiny organisms can fit on the head of a pin.

During a growing phase, bacteria increase by more than four thousand in one hour. A number of factors af-

FIGURE 7–1 Transmission of Disease-causing Bacteria from Six Sources to Humans

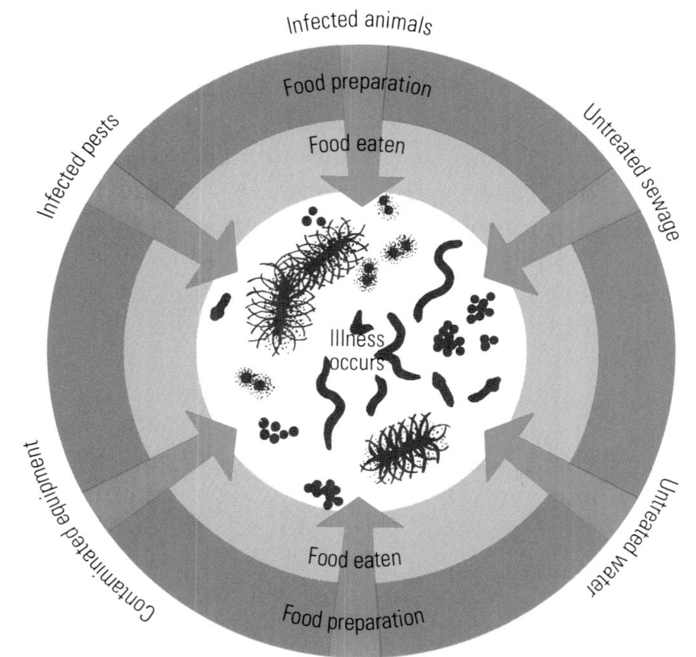

fect bacterial growth. Acid, salt, and sugar act as inhibitors and prevent the multiplication of bacteria. The presence of other factors, such as protein, moisture, temperature, and time, increase bacterial growth. The following sections discuss these bacterial growth enhancers.

Protein

Bacteria thrive in foods that contain protein, such as fresh and cooked meats, fish, poultry, eggs, milk, meat-based soups, sauces, and gravies. These foods must be stored and handled at the proper temperatures to prevent bacterial growth.

Moisture

Bacteria need moisture to grow, but they can survive in foods with low water content, such as flour, cereal, and uncooked grains. Bacterial growth is retarded in foods with high salt and sugar contents because salt and sugar remove moisture from food. This decreases the chances for bacteria to multiply. In foods with high moisture content, such as meat, poultry, fish, shellfish, dairy products, gravies, soups, sauces, and puddings, the medium for bacterial growth is excellent and the rate of bacterial multiplication increases greatly. Special care must be taken when handling these foods to protect them from contamination.

Temperature

The danger zone for bacterial growth is between 45°F to 140°F (see Figure 7–2). In this temperature range, bacteria multiply at dangerous levels. Cold foods should be stored at temperatures below 45°F, and hot foods should be kept above 140°F to prevent bacterial growth. If a food containing bacteria is frozen, bacterial growth will temporarily stop. Growth will resume, however, as the food thaws. To prevent this growth, food should be thawed while refrigerated or under cold running water. In addition, most bacteria can be destroyed when heated above 140°F. This is not true for all bacteria, however. Temperatures over 240°F are needed to destroy the spores in spore-containing bacteria. (An acid medium may also be required.) Since water boils at only 212°F, all contaminated food must be boiled for at least 15 minutes in order to destroy bacteria.

ILLUSTRATION 7–2 Bacteria need moisture to grow, but they can survive in foods with low water content, such as flour, cereal, and uncooked grains.

FIGURE 7–2 Temperature Guide for Food Safety

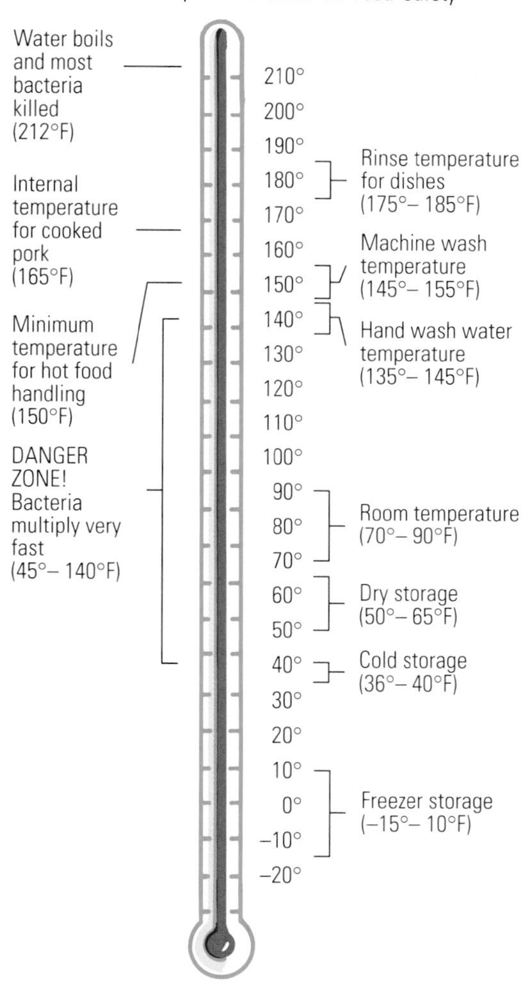

Time

The relationship between time and temperature is a key determinant of bacterial growth. Food should only be held at danger-zone temperatures (45°–140°F) for a maximum of two hours. When cooling large batches of cooked food for refrigeration or freezing, it must be passed through the danger zone as quickly as possible. This can be accomplished by using large shallow pans that help to cool food quickly. When cooling, food must reach an internal temperature of 45°F in less than two hours.

Bacterial Varieties

There are four varieties of bacteria commonly found in food. These bacteria are *Staphylococcus aureus, Clostridium perfringens, Salmonella,* and *Clostridium botulinum.*

ILLUSTRATION 7–3 Cold foods should be stored at temperatures below 45°F to prevent bacterial growth.

Staphylococcus Aureus

Staphylococcus aureus, often called staph, is a pathogenic bacterium that resembles tiny circles in the shape of grapes (see Figure 7–3). This bacterium can be found in the mouth, eyes, ears, and sinuses. It is also present on the hands and skin, especially in infected and pus-containing sores, such as cuts, burns, boils, and pimples. *Staphylococcus* grows best at temperatures between 44°F and 114°F, is highly resistant to heat, and can only be destroyed at temperatures above 165°F.

Appearance and taste should never be used as indicators for spoilage. Food contaminated with staph will look and taste normal. Staph is most often found in meats, poultry, eggs, fish, shellfish, and other protein-rich foods such as cream soups, sauces, and puddings. If susceptible food has been left out longer than two hours in temperatures between 45°F and 140°F, it should be thrown out.

Staph poisoning generally begins two to eight hours after the ingestion of contaminated food. Symptoms include sudden nausea, vomiting, diarrhea, cramps, and weakness. These symptoms usually last one to two days.

To prevent staph poisoning, food handlers must be free of infected sores and respiratory illnesses, and they should develop sanitary work habits. For example, workers should wash their hands before preparing food and pass food rapidly through the danger-zone temperatures (45°–140°F).

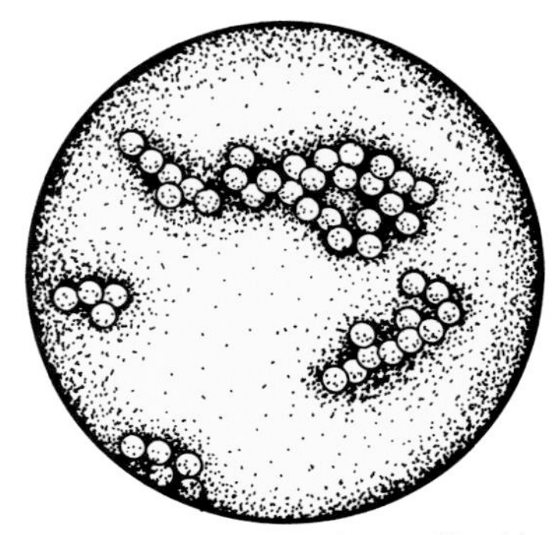

FIGURE 7–3 *Staphylococcus Aureus* (Staph)

Clostridium Perfringens

Clostridium perfringens is a spore-forming bacterium that thrives in the absence of oxygen. Besides staph, it is the most common disease-causing bacteria in the United States. *Clostridium perfringens* (illustrated in Figure 7–4) is a ball-shaped organism and grows best at temperatures between 60°F and 122°F, with optimum growing temperatures in the 109°F to 116°F range. This bacterium cannot be destroyed by boiling temperatures because of its heat-resistant spores.

Clostridium perfringens can be found in the soil and in the intestinal tracts

FIGURE 7–4 *Clostridium Perfringens*

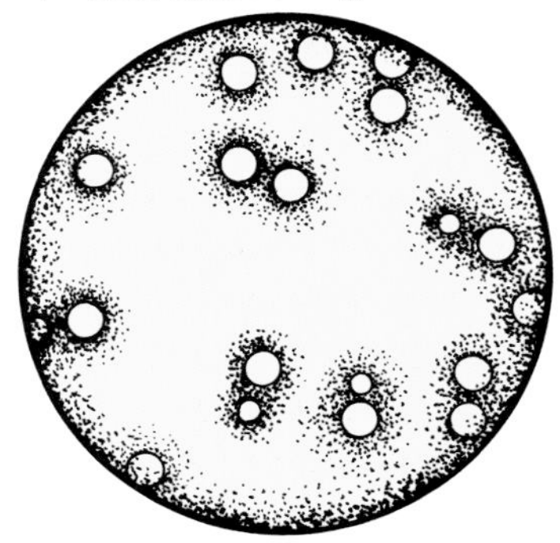

of humans and animals. Susceptible foods include raw meat, which may contain bacteria from the slaughtering process, meat gravies and sauces, and cooked meat and poultry left at room temperature for several hours.

Symptoms of *Clostridium perfringens* poisoning are similar to those that accompany staph poisoning and include nausea, vomiting, pain, and weakness. These symptoms usually appear between 8 and 22 hours after the ingestion of poisonous food. To prevent contamination from this bacteria, food service employees must be clean, and they should observe strict time and temperature controls when handling and storing food.

Salmonella

Salmonella is a rod-shaped organism with villa (see Figure 7–5). It grows best at temperatures between 44°F and 114°F and can be found in a wide range of foods, including raw meat products, poultry, eggs, dried yeast, coconut, candy, and seafood from sewage-polluted waters.

FIGURE 7–5 *Salmonella*

Salmonella does not produce toxins as do *Clostridium perfringens, Staphylococcus aureus*, and *Clostridium botulinum. Salmonella* can thus be easily destroyed by heat. This destruction can be accomplished by heating food to the boiling point (212°F).

When food contaminated with *Salmonella* is consumed, the bacteria multiplies in the victim's digestive tract and causes illness. Symptoms may include a headache, diarrhea, abdominal cramps, vomiting, and fever. These symptoms usually occur 12 to 35 hours after the ingestion of the contaminated food.

Food service managers can prevent *Salmonella* growth by taking some special precautions. Food, especially poultry, should be purchased through reputable sources that carry inspected and certified products. Insects and rodents must be strictly controlled.

Salmonella can spread easily by contaminated utensils or work surfaces. Cutting boards that have not been properly washed between uses, especially after exposure to poultry, are a frequent source of bacteria. *Salmonella* also can be spread by hands, flies, and rodents.

Clostridium Botulinum

Clostridium botulinum is a rare, but lethal, bacterium that affects the nervous system of individuals who have ingested contaminated food. This rod-shaped (see Figure 7–6), spore-forming organism's optimum temperature for growth is 95°F. Like *Clostridium*

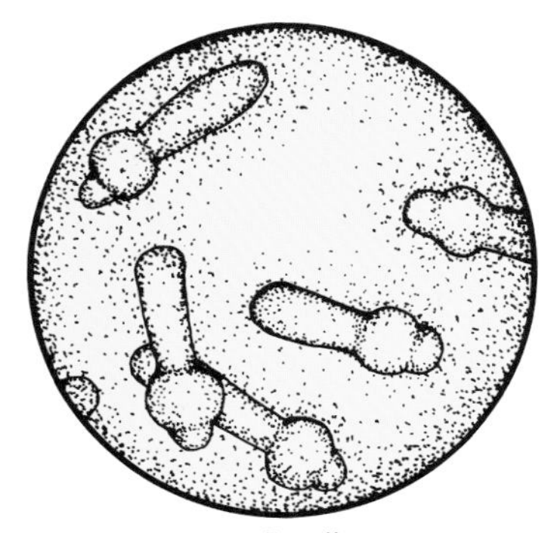

FIGURE 7–6 *Clostridium Botulinum*

perfringens, it can live and grow without oxygen.

The symptoms of botulism usually appear 12 to 35 hours after the ingestion of contaminated food and include vomiting, dizziness, double vision, and difficulty breathing. Death occurs in 50 percent of botulism cases.

Improperly processed, non-acidic canned foods, such as meats, vegetables, and vacuum-packed fish products, provide an excellent medium for bacteria growth. To prevent botulism, food service managers should discard bulging, leaky, and rusty cans. Cooks or other workers should never attempt to taste food that has an odor or color change, or if they suspect spoilage. In addition, restaurants should not serve or accept donations of home-canned products. Smoked fish must be kept refrigerated at temperatures below 38°F. *Clostridium botulinum* cannot be destroyed easily. In order to protect neighborhood pets from the deadly poison, contaminated food and its container must be boiled for 15 minutes before it is thrown away.

Parasites

A **parasite** is an organism living in or on another organism. The disease trichinosis is caused by ***Trichinella spiralis,*** a small, worm-like parasite (see Figure 7–7) frequently found in pork. Trichinosis can be transmitted to food by infected swine, rabbits, bears, rats, and other wild animals. This parasitic disease first invades the intestines and then becomes embedded in muscle tissue. Symptoms begin with nausea, vomiting, diarrhea, colic, fever, and sweating. Muscle soreness, swelling, and chills may develop. Death rarely occurs. To prevent trichinosis, all pork and pork products must be cooked thoroughly to internal temperatures above 165°F. Pork should always be served well done, with no trace of pink in the meat or juices.

FOOD CONTAMINATION

Food can be contaminated in many ways. The most common sources of contamination are chemical poisons,

FIGURE 7–7 *Trichinella Spiralis*

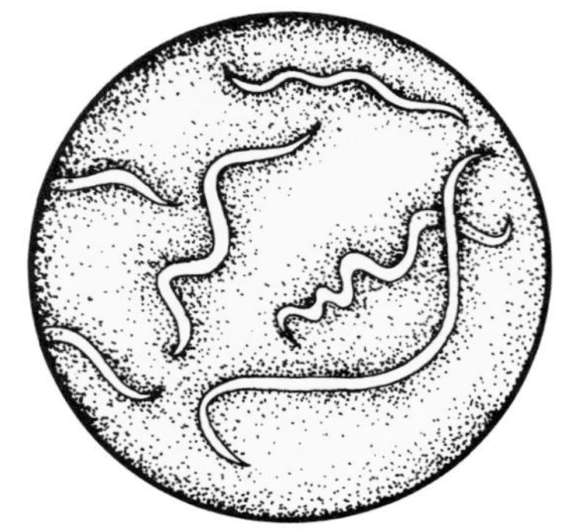

insects and rodents, infected food handlers, unsanitary equipment, improper food preparation and storage, and untreated water and sewage.

Chemical Poisons

Chemical contamination of food is a major concern of today's food service industry. Dangerous chemicals can poison food during any stage of production, from planting and processing to preparation and service. The most common chemical hazards to food are (1) **pesticides,** chemicals used to kill pests during plant growth, (2) **preservatives,** chemicals used on foods to maintain freshness and prevent spoilage, and (3) poisonous metals.

Pesticides

In recent years, American consumers have begun to voice their concern about the use of pesticides. Pesticides

NUTRITION NOTE PESTICIDES

If you were thinking of becoming a vegetarian to reduce or eliminate the risk of foodborne illness, think again. According to a 1987 report by the National Academy of Science, several pesticides used to protect growing vegetables from insects are hazardous to humans. Many of these chemicals cannot be removed by simply washing the food. Cooking vegetables destroys some of the pesticides, but it also destroys the water-soluble vitamins abundant in most vegetables. The following table lists hazardous pesticides and the foods they contaminate.

Hazardous Pesticides	Contaminated Foods
Aldicarb	Potatoes
Acephate	Bell peppers
BHC	Cabbage, sweet potatoes
Captan	Grapes
Chlordane	Potatoes
Chlorothanonil	Green beans
Dleldin	Carrots, cucumbers, potatoes, sweet potatoes, corn
DDT	Carrots
Methyl Parathion	Cantaloupe, strawberries
Lindane	Corn
Surfallate	Corn

can enter the food supply in several ways. They may be applied directly to a plant or animal in an attempt to protect it from insects. Plants and animals may consume poisonous agents during their growth, thus accumulating harmful substances in their living cells. Food can also become contaminated by chemical agents during processing or preparation.

Food service managers do not have much control over the sources of their food, besides choosing reputable suppliers. They can, however, prevent contamination of food in their establishments by properly training their employees. Workers should label and store poisonous chemicals, in their original containers, away from all food products.

Preservatives

The effects of long-term consumption of preservatives, or additives, have not yet been determined. Preservatives are still a subject of debate among scientists and legislators. **Sodium nitrite,** for example, is an additive used in lunch meats, bacon, hot dogs, and other products. Nitrates serve an important purpose—they ensure longer shelf life and combat botulism in processed foods. Their effects on humans, however, are unclear. Most experts thus recommend that a person eat a variety of foods to prevent a high intake of any particular preservative.

Some preservatives clearly have a bad effect on people. Monosodium glutamate (MSG), a commonly used flavor enhancer, can cause illness if consumed in excess. Symptoms of MSG poisoning include flushing of the face, dizziness, headaches, and nausea. These symptoms have sometimes been called "Chinese restaurant syndrome" because most Chinese recipes contain large amounts of monosodium glutamate.

Poisonous Metals

Poisonous metals are another source of food contamination. This type of contamination occurs through the interactions of a particular food with metal, oxygen, and light. When **galvanized** (zinc-coated) copper or brass containers are used to cook or prepare a high-acid food, such as tomatoes, lemonade, fruit gelatins, or fruit punches, the coating reacts with the acid, contaminating the food. Gray enamelware containers plated with antimony, cadmium, or lead glaze can also react with food and beverages. To prevent metal poisoning, food service establishments should use stainless steel, glass, or plastic cookware and utensils.

Insects and Rodents

Food attracts rodents and insects, including rats, flies, roaches, and ants. These pests carry disease-causing microorganisms and contaminate any food they touch. In addition, insects and rodents cause more than $10 billion in food spoilage and waste each year. This amount includes fires that

ILLUSTRATION 7–4 To prevent metal poisoning, food service establishments should use stainless steel, glass, or plastic cookware and utensils.

are started by rats gnawing through electrical wiring.

Cleanliness is the key to avoiding insect and rodent contamination. Pests can be kept away from a food service establishment by eliminating their breeding places, blocking entryways, covering food containers and trash cans, keeping the food operation area clean, and installing screens in doors and windows.

Food Handlers

A food service operation's success depends on the health and grooming habits of its employees. Food service employees can easily transmit disease-causing bacteria to customers, either directly or indirectly through food contamination. Managers must have an established policy requiring physical examinations for new employees and annual examinations for all others. Employees should be instructed to report illnesses or injuries such as colds, sores, wounds, or lesions as soon as they appear. They should not be permitted to work when exhibiting any of these symptoms.

An employee with an infected burn or sore on her or his hand should not work. There can be exceptions to this rule, however, depending on the work situation. If cut or burned employees are allowed to work, the affected area should be bandaged carefully and covered with a plastic glove.

Employees should not wear rings and other jewelry that collect dirt or that could fall into food during preparation or service. Fingernails must be kept short and clean. All food service workers should get in the habit of washing their hands with soap following any act that could possibly transmit bacteria.

Restaurants often receive complaints from customers about hair in their food or on their table. Food service managers should strictly enforce regulations that require all employees engaged in food preparation and service to wear hair nets, head bands, caps, or other effective hair restraints.

Food service establishments are judged by first impressions. If a customer's first impression is negative, he or she may never return. As discussed in Chapter 6, the appearance and attitude of employees can often make or break a business.

Equipment Sanitation

Properly cleaning hand tools, dishes, glassware, and utensils helps eliminate disease-causing bacteria that could be transmitted through dirty utensils and serviceware. Mechanical dishwashers are efficient, sanitary, and save on labor costs.

Clean equipment helps a food service establishment meet the standards of its local health departments and the National Sanitation Foundation. In addition, cleanliness helps prevent the breakdowns and corrosion that plague neglected equipment. Food service employees should follow a regular cleaning schedule. It is the manager's responsibility to see that employees do their jobs properly and maintain a sanitary environment.

Food Storage

All foods should be stored properly to prevent spoilage. They should also be grouped in an organized manner to allow for easy access. The first and most important storage principle to remember is **FIFO,** "first in, first out." FIFO means that new stock should be placed on the back of shelves so that older stock will be used first. If this is not done, products pushed back on the shelf will eventually spoil.

Generally, food service establishments use three types of storage: refrigerator, freezer, and dry. Canned products should be stored separately from dry goods.

Perishable foods, such as cheese, milk, eggs, fresh fruits, and vegetables, should be kept under refrigeration at temperatures between 34°F and 40°F. Meats, fish, and poultry may also be kept at this temperature, but their shelf life is very short. It is best to store these items in a freezer if they are not used in the first two to three days after purchasing. Freezer temperatures should range between −15°F and 10°F.

Leftover food is susceptible to microorganisms and should be used as soon as possible or stored in a freezer. All refrigerated and frozen foods must be placed in airtight containers to prevent food from drying out and absorbing odors given off by other foods.

Dry storage areas should be well ventilated and the temperature should range between 50°F and 70°F. Food should be held in tightly covered con-

ILLUSTRATION 7–5 Perishable foods, such as cheese, milk, eggs, fresh fruits, and vegetables, should be kept under refrigeration at temperatures between 34°F and 40°F.

MATH SCREEN

The cooling time of hot foods depends on the depth of the container in which they sit.

Formula: Cooling time (in hours) = (½ depth)^2

Container depth sizes: 2 inches
4 inches
6 inches
8 inches

What is the cooling time of a 2-inch-deep container of hot food?

$(1)^2 = 1$ Cooling time is one hour.

What is the cooling time of a 4-inch-deep container of hot food?

$(2)^2 = 4$ Cooling time is four hours.

Test your math skills: What is the cooling time of a 6-inch-deep container of hot food? An 8-inch-deep container?

tainers well above the floor; the lowest shelf should be at least 18 inches above the floor. In addition, open-slatted shelves allow an even airflow around foods to prevent spoilage and insect infestation.

SUMMARY

Each year thousands of cases of foodborne illness are reported in the United States. Disease-producing bacteria can be found everywhere, and they grow rapidly when conditions (such as protein availability, moisture, temperature, and time) are right.

Food service employees need to be familiar with the four varieties of bacteria commonly found in food—*Staphylococcus aureus, Clostridium perfringens, Salmonella,* and *Clostridium botulinum*—and how to prevent their growth. Bacteria can be transmitted to food and water by insects, animals, and humans. In addition, parasites, pesticides, preservatives, and poisonous metals can also contaminate food. Thus the principal causes of food contamination are infected food handlers, contaminated food supplies, unsafe food storage, unsanitary equipment, and hazardous chemicals.

Regular visits from local, state, and federal inspectors help ensure that the food an establishment serves is safe for human consumption. Ultimately, however, food service managers are responsible for public safety. They must follow their local health department's written guidelines to maintain sanitary standards and eliminate the threat of foodborne illness.

QUESTIONS

1. What are the four federal agencies responsible for sanitation standards and regulations? List the major responsibil-

ities of each of these agencies.
2. What are four bacteria that cause foodborne illnesses? List the symptoms associated with each illness.
3. What are four sources of food contamination?
4. What are four conditions that favor bacterial growth?
5. Where are *Staphylococcus aureus* bacteria found?
6. What are three types of chemical food poisons? How can they be prevented?
7. How can food service operations stop rodents and insects from contaminating food?
8. What is the appropriate temperature range for refrigerated food?
9. What is the appropriate temperature range for frozen food?
10. What is the appropriate temperature range for dry food storage?

ACTIVITIES

Activity One Invite a guest speaker from your town or city's health department to inspect your classroom kitchen or food preparation area for cleanliness and sanitation. Ask the official to discuss any violations and ways to correct them. Allow at least 15 minutes during the visit for questions and answers. Make sure that you and your fellow students have each prepared at least one question to ask the official.

Activity Two Bacteria can be found everywhere, even on things that appear clean. If you look at the world under a microscope, you will find that you are surrounded by different types of bacteria.

For this activity you need to have the following:

1. A microscope (may be borrowed from your school's library or biology classroom)
2. Four agar petri dishes (standard count or typtic soy agar with sheep's blood)
3. A lot of germs!

Label the petri dishes and contaminate each dish with germs from the following sources:

- A cough or a sneeze
- A skin cut or a pimple
- A hair
- A table, desk, or counter

Germs should be gathered by scraping your skin, hair, table, and so on, with a clean toothpick. Place the contaminated petri dishes in a dark place where the temperature is between 75°F and 100°F. Store for 24 hours. Examine the petri dishes under a microscope and identify the type of bacteria growing. Place the dishes back in storage and monitor them daily. Which bacteria grows the fastest?

8 SAFETY

VOCABULARY

Occupational Safety and Health Administration (OSHA)
health inspectors
noncompliance
Class A fires
Class B fires
Class C fires
fire extinguishers
grounded equipment

OBJECTIVES

After studying this chapter you should be able to do the following:

- Describe the functions of OSHA
- Explain the five most common injuries that occur in food service kitchens and several ways to prevent each one
- Identify three classes of fires and the types of fire extinguishers used to extinguish them
- List several ways to prevent food service fires
- Demonstrate the ability to use a fire extinguisher
- Recognize a food service manager's liability if guests are injured

The operation of a food service facility creates many potentially dangerous situations that threaten the safety of workers and guests. Minor injuries from burns, cuts, falls, strains, and shocks are common. The type of equipment used, the high temperatures required in food and beverage preparation, and the speed at which materials are handled make safety a major concern for food service employers and managers. Approximately 85 percent of all accidents result from human error and about 15 percent are caused by faulty equipment and other unsafe conditions. Inexperience and inadequate training are the major reasons for accidents resulting from human error. All types of food service establishments should have an accident prevention training program to help eliminate employee and guest accidents.

SAFETY ENFORCEMENT

Food service employers are ultimately responsible for preventing injuries in their establishments. Under the 1970 Federal Occupational Safety Administration Act, employers are required to furnish a place of employment free from known hazards. The **Occupational Safety and Health Administration (OSHA)** was created within the Department of Labor to

- encourage employers and employees to prevent hazards in the workplace and to improve existing safety and health programs;
- establish reporting and record keeping procedures to monitor job-related injuries and illnesses;
- develop job safety and health requirements and effectively enforce them; and
- encourage each establishment to have someone on staff trained in first aid and readily available during work hours.

As discussed in Chapter 7, **health inspectors** working for local health departments make regular visits to food service establishments. In addition to inspecting for the sanitary handling of food, they check to see that an establishment has eliminated safety hazards from the workplace. **Noncompliance,** which means that a business has not followed safety regulations, can result in penalties that range up from a $10,000 fine to six months in jail for an employer.

Employees working in food service establishments are surrounded by dangerous energy sources. These sources include steam, electricity, and gas. They can cause burns, shocks, and injuries from explosions. Poorly designed kitchens can contribute to accidents caused by cross-traffic problems, dangerous placement of equipment, and narrow aisles. Adequate aisle space should be provided to prevent workers from colliding while carrying trays or other items.

Accidental injuries can also involve guests. Injuries can result from food served at extremely high temperatures or from falls on wet or slippery floors.

Experience indicates that preventing accidents in the workplace eliminates painful injuries, decreases anxiety among workers, and cuts down on costs. Frequent accidents result in increased insurance rates, damaged materials, and lost employee work time. In some cases, new employees must be hired and trained until injured employees are able to return to work. All these factors may contribute to low morale, low productivity, and, ultimately, low profits.

COMMON TYPES OF INJURIES

Every food service establishment must have a set of safety rules to protect its employees. These rules should include guidelines for appropriate dress, the proper use of equipment, and the prevention of accidents and fires. Effective safety practices help to prevent the most common injuries that occur in food service facilities: falls, cuts, burns, and strains. Injuries from electrical shocks and explosions occur less frequently, but they can be serious.

Falls

Falls are the number one cause of accidents in food service kitchens. Most often, falls are related to or caused by poor lighting or slippery or wet floors. One way to prevent falls is to make sure that all employees wear appropriate shoes. Generally, food service employees spend many hours on their feet. Comfortable shoes contribute to better performance and safety. Flat, rubber-soled, leather shoes are best to prevent slipping and to ensure comfort.

Other ways to prevent falls are to properly clean floors so that they are free of grease, to mop up excess moisture on humid days, and to immediately take care of spills and items that have fallen on the floor. If necessary, managers should sprinkle salt on the floor to prevent slipping.

Hallways must be well lighted. Food service establishments should have a rule that employees warn other employees when walking behind them. This rule prevents sudden movements that may result in injuries. Swinging doors leading into and out of the kitchen should be clearly marked *Entrance* or *Exit* on both sides and should be used as marked.

ILLUSTRATION 8–1 Wet floors must be clearly marked to prevent falls.

Cuts

Cuts are the second most common injuries in food service kitchens and usually result from the improper use of knives or other equipment. Handling knives correctly and practicing all safety precautions that come in writing with every piece of equipment help to eliminate most of these accidents.

Cutting is easier and safer with a sharp knife; with a dull knife, a worker needs to apply pressure that might cause the knife to slip. The use of cutting boards or wood tables (butcher blocks) prevents the dulling of knives. Knives should be used for their designated purpose only and not for opening cans or jars. When cutting, chopping, or slicing, workers should hold the knife firmly down on the cutting board and slice away from the body. Knives must always be carried by their handles with the tips pointed down. They should never be stored loosely in a drawer but rather on a knife block, which is safer and helps keep the edges sharp.

When employees handle knives, they must pay very close attention to what they are doing. This is not the time to play around. Employees should be instructed that, if a knife falls during handling, they should *never grab at a falling knife.* They should get out of the way and let it fall. In addition, it is a good policy to always clean knives separately.

The improper use of equipment is a major cause of cuts. All equipment should be operated by trained employees. When cleaning equipment, a good rule is to disconnect the power plug and then proceed with the cleaning. This will prevent accidentally pushing the *on* button while cleaning. After cleaning equipment, employees should make sure that the unit is off before reconnecting it to a power outlet.

Mixing equipment should never be started without the bowl and beaters in their proper places and securely fastened. Finally, employees should use a wooden stomper, rather than their hands, when feeding meat or other items into a grinder.

Burns

Due to the nature of work in food service establishments, burns are the most serious hazard. Employees must be well trained to cope with all possible sources and kinds of fires. They should know the location of fire extinguishers and the appropriate actions to take in case of fire. In addition, it is important that all employees know where the telephone numbers for the fire department and other emergency services are posted.

The first few minutes of a fire are critical. The initial actions taken can determine whether a fire is smothered quickly or causes extensive damage or painful injuries to employees or guests.

All fires need air, fuel, and heat to ignite. If one of these ingredients is

ILLUSTRATION 8–2 Knives should never be stored loosely in a drawer but rather on a knife block, which is safer and helps keep the edges sharp.

missing, a fire will not start. As a fire burns, it uses air; thus, if its oxygen supply is cut off, a fire will stop. Fires are often fueled by grease or cloth, which are both commonly found in food service kitchens. As with its air supply, if a fire's fuel is removed, it will stop burning. Once a fire has started, however, it is difficult to remove the third ingredient, heat, because a fire creates its own heat as long as it has fuel and air.

Most fires can be categorized into three classes: Class A, Class B, and Class C. **Class A fires** involve combustible materials, such as wood, paper, and cloth. **Class B fires** are all those that are fueled by grease or oil. **Class C fires** are electrical. When fires occur in food service kitchens, they are most often Class B and Class C fires because of the equipment and materials found there.

To smother a Class B or Class C fire, a person must cut off its oxygen. Putting out a Class A fire requires the removal of its fuel or the cooling of its heat. Food service facilities thus need several types of **fire extinguishers.** Multipurpose dry chemical extinguishers are the best type to have available, because they can be used to fight all three classes of fires. These fire extinguishers are labeled *ABC*.

Carbon dioxide gas extinguishers smother the fire without conducting electricity. In addition, the gas is extremely cold and can lower a fire's heat level without damaging equipment. Figure 8–1 (page 96) shows the different types of fire extinguishers, and Figure 8–2 (page 97) illustrates how to properly use an extinguisher.

Fryers, broilers, and range tops are the most common areas where grease fires occur. These areas, including the venting hoods above them, should be cleaned biweekly to prevent the accumulation of grease that often leads to fires. When using frying equipment, food service employees should never throw wet, cold food or any liquid other than grease into a hot grease fryer. Doing so is dangerous; hot grease can splatter in an area a few feet in diameter, causing severe burns or starting fires.

Cooks should keep a large box of baking soda easily accessible to use in case of minor grease fires on the range top or in the broiler. If baking soda is not available, salt can also be used to smother a fire. Table 8–1 summarizes the classes of fires, materials involved, methods of prevention, and ways to extinguish them.

Many burns occur from carelessness and can be easily prevented. Employees should always use a dry towel

FIGURE 8–1 Types of Fire Extinguishers

KIND OF FIRE		APPROVED TYPE OF EXTINGUISHER						
Decide the class of fire you are fighting...	...then check the columns to the right of that class	Match up proper extinguisher with class of fire shown at left						
		Foam (solution of aluminum sulfate and bicarbonate of soda)	Carbon Dioxide (carbon dioxide gas under pressure)	Soda Acid (bicarbonate of soda and sulfuric acid)	Pump Tank (plain water)	Gas Cartridge (water expelled by carbon dioxide gas)	Multi-purpose Dry Chemical	Ordinary Dry Chemical
A	Class A Fires Use these extinguishers Ordinary combustibles • wood • paper • cloth etc.	A B	X	A	A	A	A B C	X
B	Class B Fires Use these extinguishers Flammable liquids, grease • gasoline • paints • oils etc.	A B	B C	X	X	X	A B C	B C
C	Class C Fires Use these extinguishers Electrical equipment • motors • switches etc.	X	B C	X	X	X	A B C	B C

Source: Adapted from a National Institute of Occupational Safety and Health chart.

1. Hold upright, pull ring pin.
2. Start back from ten feet distance away from the fire.
3. Aim at the base of the fire and squeeze the lever.
4. Move in a sweeping motion, first at the base of the fire and gradually sweeping forward and upward.

FIGURE 8–2 How to Use a Fire Extinguisher

or pot holder when handling hot utensils. Wet or damp cloths conduct heat quickly. Cooks can avoid burns to the face or neck by slowly lifting pot lids away from their bodies to allow steam to escape. Pot and skillet handles should be pushed away from the edge and toward the center of a range to prevent bumping or knocking over hot pans.

It is important that all employees know the location of the fire exits. An evacuation procedure must be posted and practiced on a regular basis.

TABLE 8–1 Fire Safety

CLASS OF FIRE	CAUSES	PREVENTION METHODS	EXTINGUISHING METHODS
A	Wood, paper, or cloth too close to heat	Store materials away from heat	Use water; smother with damp blanket or use a Class A fire extinguisher
B	Grease (oil or fat) too close to heat	Use appropriate size pans to prevent grease spills; do not overheat grease; handle grease carefully so that it doesn't splatter; clean exhaust fan	Smother with blanket; smother with baking soda; use Class B fire extinguisher
C	Faulty wiring leading to electrical sparks	Make sure all electrical wiring is safely installed; turn off the equipment power if you smell a sharp and burning odor	Use Class C fire extinguisher

ILLUSTRATION 8–3 Cooks can avoid burns to the face or neck by slowly lifting pot lids away from their bodies to allow steam to escape.

Strains

Strains can be avoided by properly handling heavy objects. Figure 8–3 illustrates proper lifting techniques, which include using the leg muscles, keeping the back straight, bending the knees, and pushing upward with the thigh muscles while lifting. Heavy equipment or boxes that do not require lifting should be moved with a cart or a dolly. Equipment must be installed in a stable position so that it does not move during food preparation or cleaning.

Electrical Shocks and Explosions

Electrical shocks and explosions are less frequent than falls, cuts, burns,

FIGURE 8–3 The Proper Way to Lift Heavy Objects

1. Squat down to the level of the load with knees bent and back as straight as possible. Do **not** bend over from the waist. This may strain the back and cause injury.
2. Use the leg muscles while pulling in the stomach muscles.
3. Slowly stand up without sudden movements that might cause injury.

ALCOHOL AWARENESS

Alcohol (sometimes called ethyl alcohol or ethanol) is a small molecule that is completely soluble in water. Like other drugs taken orally, alcohol is absorbed mostly by the stomach and upper small intestine. How fast it is absorbed depends on the amount of food and water present in the stomach at the time of consumption. Most alcohol is carried in body water; thus blood-alcohol content (the amount of alcohol in the blood) is used as the basis for calculating how much alcohol is actually in a person's tissues.

Beverages containing alcohol are widely used and their abuse has become a major problem in the United States. Driving while intoxicated, for example, is a prime cause of automobile accidents. The effects of alcohol abuse are divided into two categories: acute reactions are those that result shortly after the intake of alcohol, and chronic reactions are the long-term effects of alcohol. Alcoholism is the compulsive or excessive use of alcohol over a long period of time. The effects of alcohol depend on the amount the person drinks, how often the person drinks, and the alcohol content in the beverage.

The employer or manager of a food service establishment that serves alcoholic beverages should be familiar with the short-term as well as the long-term effects of alcohol abuse. They should also know techniques to discourage guests from drinking more alcohol than their bodies can handle. All bartenders and servers should be educated to recognize when a guest has had too much to drink and how to offer alternative beverages. It is important that a guest leaves capable of driving safely; otherwise, an establishment's staff should help the guest call a cab or friend, or help obtain another form of transportation.

and strains, but they can cause serious injuries when they occur. Electrical equipment can be dangerous to work with if it is not properly grounded. **Grounded equipment** has a three-pronged plug that allows electricity to travel to and connect with the ground instead of with a person operating the equipment.

Even with grounded equipment, however, precautions must be taken to prevent a shock. Electricity travels through water quickly. Employees should make sure that their hands are dry when plugging in, unplugging, or operating electrical equipment. Electrical controls and wires are not designed to get wet, and they must be kept dry at all times. In addition, all cords should be inspected often for worn sections.

All gas equipment must be in good

MATH SCREEN

Alcoholic beverages are divided into three categories: wine, beer, and distilled liquors (such as whiskey, bourbon, or rum). These types of beverages are different in caloric content, taste, water content, and alcohol content.

The following examples show how to determine the number of grams of alcohol in different drinks. You can figure grams per drink with this formula: milliliters of a beverage × percentage of alcohol = grams of alcohol.

- Beer: A 12 oz beer is approximately 3.8 percent alcohol and thus has 13.4 g of alcohol.

  (30 ml/oz × 12 oz = 360 ml × 0.038 = 13.4 g of alcohol).

- Wine: A typical glass of wine is 11 percent alcohol (and may be higher depending on the wine). A 5 oz glass of wine thus contains 16.5 g of alcohol.

  (30 ml/oz × 5 oz = 150 ml × .11 = 16.5 g of alcohol).

- Distilled spirits: Distilled spirits, like brandy or rum, are measured in proof degrees. 1 proof equals .5 percent alcohol content; thus, 80 proof whiskey contains 40 percent alcohol. Some distilled beverages range from 35 to 50 percent alcohol content or 70 to 100 proof. A standard measure for distilled beverages is 1.5 oz (45 ml). Thus, an 80 proof shot would contain 18 g of alcohol.

  (45 ml × 0.40 = 18 g).

Test your math skills: How many grams of alcohol are in an 8 oz glass of wine?

working condition and should include an automatic safety device that prevents explosions by turning off the gas if the equipment's pilot light blows out or is not lit. If a gas leak occurs, the main gas control must be turned off, the room or rooms must be vented, and the proper authorities should be called immediately for repairs. Natural gas itself is odorless, but most gas companies add a strong, offensive odor to their gas so that leaks can be easily detected.

Food service establishments contain a number of serious safety hazards. Good safety practices must be consistently observed to ensure the safety and well-being of employees. If they are not, employee complaints can initiate a government inspection that could result in penalties for a food service operator. Frequent accidents also cause insurance rates for medical expenses, liability, and worker's compensation for injured employees to skyrocket. In addition, numerous accidents in a food service facility can adversely impact employees' morale and seriously erode their confidence in management practices and policies.

FRONT-OF-THE-HOUSE SAFETY

Guest safety is as important as employee safety. The owner of a food service facility is responsible for all accidents that occur in his or her establishment, even those accidents caused by employee carelessness. Guest injuries can damage an establishment's reputation and can result in fines or imprisonment. These consequences can be avoided by a well-designed facility and properly trained employees. A constantly reinforced safety program for all food service employees helps to ensure the safety of guests. Employees must be trained to balance and carry trays carefully, to serve hot foods and liquids safely, and to mark wet floors properly.

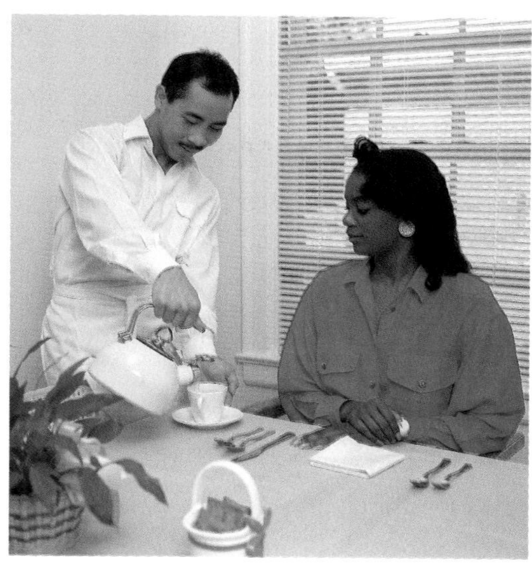

ILLUSTRATION 8–4 A constantly reinforced safety program for all food service employees (including instruction on how to pour hot liquids carefully) helps to ensure the safety of guests.

NUTRITION NOTE ALCOHOL ABUSE AND ITS EFFECTS ON NUTRIENTS

Frequent alcohol consumption interferes with the body's ability to regulate food intake, absorption, metabolism, and nutrient storage. Ultimately, the damage that alcohol does to live cells leads to malnutrition. Protein, B vitamins, and fat soluble vitamins, as well as some minerals, are usually deficient in people who abuse alcohol.

Calories from food and drink are a major source of the body's energy and nutrients. The calories found in alcohol, however, do not contain any nutrients. When a person drinks a lot of alcohol, his or her hunger is controlled by the alcohol's calories, and she or he eats less food. After a period of time, many of the people who abuse alcohol thus become nutrient deficient.

Abusers of alcohol should

- seek professional help in an effort to stop drinking;
- eat plenty of nutritious foods, such as meats, fruits, vegetables, dairy products, and cereals, daily; and
- see a doctor or dietitian for advice on nutrient supplements.

CAREER PROFILE
POSITION: FOOD SERVICE SUPERVISOR

Qualifications:

- Must have a two-year associate's degree in food service management or business or a two-year associate's degree in dietetics plus a minimum of one year of experience in food service
- Must have good verbal and written communication skills and the ability to lead

Responsibilities:

- Supervises food service employees working in restaurants, hotels, schools, colleges, universities, or similar food service operations
- Trains employees and coordinates work schedules for the efficient use of resources
- Keeps records on the number of meals served and their costs
- Oversees the proper cleaning of the kitchen, dining area, equipment, and utensils
- Makes sure that employees follow safety and sanitation guidelines at all times
- Maintains inventory stock levels
- May assist other managerial staff members with menu planning and the recruitment and hiring of new employees

SUMMARY

The Occupational Safety and Health Administration (OSHA) is a federal agency that requires food service employers to comply with safety standards that protect employees and guests. These standards are designed to prevent accidents, such as falls, cuts, burns, strains, and electrical shocks, which are a constant danger because of the powerful machinery, cutting utensils, and high-temperature cooking equipment involved in food service.

Well-trained workers can often avoid accidents; a large number of employees, however, know little about safety regulations and procedures. Thus a good training program is vital, and it should be given to each employee immediately after hiring. Periodic training sessions and evaluations should be conducted to further promote a safe food service environment.

QUESTIONS

1. What is OSHA?
2. Why does a food service employee need to be familiar with OSHA?
3. What are four regulations regarding food service established by OSHA?
4. What are four of the most common hazards found in food service kitchens?
5. What are three ways to prevent falls?
6. What are three ways to prevent cuts?
7. What are four ways to prevent burns?
8. What are two ways to prevent strains?

9. What are two ways to prevent electrical shocks?
10. What problems might a food service establishment owner face if a guest is injured?
11. What are the three classes of fires? What type of extinguisher is used for each?

ACTIVITIES

Activity One Invite a speaker from your town or city's fire department to explain the use of different fire extinguishers and to inspect the kitchen in your classroom or food preparation area for fire safety. Permit time for classroom discussion. You should also practice using the different types of fire extinguishers.

Activity Two Visit a restaurant in your city or town and request permission from the owner or manager to inquire about the facility's fire prevention methods. Write a short paper on the inquiries and observations. While visiting the restaurant, you should pay special attention to the following:

- Floors
- Hoods over the cooking areas
- Electrical outlets
- Paper and cloth products close to fire sources
- Fire extinguishing equipment and fire alarms
- Emergency exits
- Markings on doorways
- Plans for emergency evacuation

PART FOUR

THE PHYSICAL PLANT

9
FACILITY LAYOUT AND DESIGN

VOCABULARY

cook-chill
cook-freeze
layout
masonry
fluorescent
incandescent
circuit breaker
wattage
volt
ampere
braziers

OBJECTIVES

After studying this chapter you should be able to do the following:

- State the common goals of food service facilities
- Outline a basic food service layout plan
- Identify several types of materials used in building construction
- Discuss the factors that affect facility layout and design
- Discuss the necessary layout components for a fast-food outlet

The past 30 years have brought many food service innovations. These innovations include central kitchens, **cook-chill** and **cook-freeze** technology (which involves prepackaged, portioned convenience foods that are ready to eat with minimal preparation), back-to-scratch cooking, and on- and off-premises catering. As a result, the layout and design of food service facilities have changed. A **layout** is a plan that shows a facility's physical set-up, including construction features and the locations and space allowances for different work stations and equipment. The success of a design will depend upon the designer's knowledge and understanding of recent innovations. Through careful research on probable costs and operational needs, a good food service planning team can develop an efficient and flexible layout that will allow an organization to accomplish its work and incorporate future changes.

PLANNING A FACILITY'S LAYOUT AND DESIGN

In order to build an appropriate food service facility, an owner or manager should determine her or his goals. After these goals are prioritized, the owner or manager needs to organize a planning team.

Goals

The major goals common to food service owners, directors, and managers are as follows: (1) efficient and high-quality production, preparation, and service, (2) attractively displayed food, and (3) satisfied customers. Preparing a list of questions dealing with customer tastes, habits, convenience, and buying ability can help determine whether a facility will meet these goals. The following are examples of questions most frequently used to survey a business's potential clientele in order to forecast goal outcome:

- Who are the customers?
- What are their occupations? incomes? age groups?
- What is the predicted number to be served daily?
- What is the best time for maximum service?
- What types of foods are most appealing to customers?
- What is the ethnic background of the majority of the customers?
- How will customers' religious beliefs affect their food choices?
- Can the operational expenses be supported? How?

After all the questions have been answered and a menu developed, the owner or manager must put together a planning team of experts to help construct a master plan for layout and design.

The Planning Team

A complete planning team, which includes an architect, building contractor, engineer, and a food service equipment expert, can assist a food service director with the physical and operational characteristics of his or her facility. These characteristics include building design and materials as well as space and kitchen layout. Decisions about building design and materials determine the type of floors, walls, ceilings, lighting, and plumbing a facility has. Space and kitchen layout considerations focus on the basic areas and stations needed for food preparation. The size of different areas and the materials used to construct them affects costs, work productivity, ease of cleaning, and safety and sanitation features.

ALCOHOL AWARENESS

Part of a food service manager's responsibility is the safety of guests. If a food service establishment serves alcoholic beverages, both employer and employees are responsible for making sure that guests can drive home safely. The bar area should be designed to include equipment such as an automatic coffee maker and nonalcoholic beverage dispensers, as well as appropriate mixers and blenders for nonalcoholic drinks. Arrangements to include this equipment should be made during the planning stages. Waiters, waitresses, and bartenders who serve alcoholic beverages should be fully aware of their responsibilities and should monitor alcohol consumption closely.

The following guidelines should be used to protect guests from drinking-and-driving accidents:

1. Do not serve alcoholic beverages to guests who have already had too much to drink.
2. Do not encourage the overconsumption of alcohol.
3. If possible, take car keys away from a guest who has had too much to drink, and call a family member or a friend.
4. If no family members or friends can drive an intoxicated guest home, call a cab.
5. As a last resort, offer to drive an intoxicated guest home yourself.
6. The best rule is to practice moderation and encourage others to do so.

BUILDING DESIGN AND MATERIALS

New developments in materials have strongly influenced architectural designs during the past decade. Stronger, more versatile materials made of metal, glass, and masonry now allow for durable, attractive building designs. **Masonry** includes brickwork or stonework. Geographic location and weather strongly influence choices in materials. Extreme changes in temperature cause some materials to expand and others to shrink, which can damage a building. Other materials do not insulate a facility from extremely cold or hot temperatures.

Brick, stone, tile, marble, cement, metal, and wood are materials used to construct buildings. A material's toughness, hardness, and resistance are important considerations in determining which materials are best suited for a job. Wood is least likely to be chosen for a food service facility because it is extremely flammable.

Interior and exterior design and decorative colors are influenced by different trends. Often, designs are repeated from past decades. In the mid-to late 1980s, for example, popular decor and architectural designs were influenced by the styles of the 1950s, with buildings being designed in "modern" square shapes. Pale pinks, light blues, mauves, and forest green were the colors of choice in these buildings. Other popular designs of the 1980s included southwestern, country, and Spanish architecture and decor.

Walls and Ceilings

Walls and ceilings in food service facilities must meet a number of criteria. Walls should fit securely against the floors, leaving no gap, and they should be covered with some type of base. Rounded baseboard corners help to prevent the accumulation of dirt and permit ease of cleaning. Cooling and heating radiators, electrical wiring, plumbing, pipes, and ventilator faces should be hidden inside the walls or ceilings.

The amount of light and sound that penetrates or reflects off walls is determined by their texture and color. Texture and color must be chosen carefully, depending on the needs of a particular room or area. Light colors, for example, tend to make an area seem larger. Walls and ceilings should always have a sanitary and pleasing appearance for customers. Enamel paint is inexpensive and often preferable to other types because it is easy to clean and can be repainted as often as necessary.

Glazed tile is the best material to use in a kitchen, especially on the walls behind food preparation areas. Glazed tile is almost always installed from six to eight feet high behind grills, stoves, fryers, and food preparation areas. Once installed, glazed tile is durable and easy to clean.

ILLUSTRATION 9–1 Popular restaurant designs of the 1980s included country architecture and decor.

ILLUSTRATION 9–2 Spanish architecture and decor was a common restaurant design during the 1980s.

Lighting and Wiring

Food looks more appealing in daylight than in artificial light. Thus it is advisable to provide as much natural light as possible. Natural light also lowers costs and increases worker productivity. Research shows a positive psychological effect on employees, as well as on guests, when they can look outside while working or eating.

When natural light is unavailable, two types of artificial lights are commonly used in food service facilities: fluorescent and incandescent. **Fluorescent** lights, which are glass tubes coated on the inside with a substance that gives off light, are available in a variety of warm colors. **Incandescent** light is produced by a filament of conducting material contained in a vacuum and heated by an electrical current. Incandescent lamps generate more heat and do not burn as long as fluorescent lamps. Choosing the right light and color can enhance the appearance of food. A yellow-white incandescent light is preferred for food displays placed in a dining area. Whatever choices are made—natural or artificial, fluorescent or incandescent—the amount of light is important. Suitable lighting reduces eye strain and enables employees to work better, thus increasing productivity.

Adequate wiring is critical at a food service facility. Electrical power must be distributed evenly to avoid overloading or a loss of energy. All electrical wiring, panel boards, switches, and circuit breakers should comply with the regulations and requirements of the National Electrical Code. A **circuit breaker** is a device that automatically interrupts an electrical current when the flow of energy becomes excessive. In addition, all power equipment must be compatible with the power supply to promote the even distribution and the efficient use of energy. To be sure that a facility has enough power available, each piece of equipment's **wattage**—the amount of energy it needs to work—should be included in the original plan prepared by an engineer or food service manager equipment specialist. Food service equipment generally uses 110 volts or 220 volts of electricity. A **volt** is a unit of electrical potential or potential difference. One volt is the electromotive force that will cause a current of one **ampere,** a unit of electrical energy, to flow through a conductor whose resistance is one ohm. (For those who wish to pursue it, more detailed information on the fascinating area of electrical units of measure is available from the public library.)

A facility's electrical fixture designs should match its overall architectural decor. All lights should be positioned so that, when employees or objects are in front them, important working areas are not shadowed.

Plumbing

Plumbing is another key consideration for building design. An adequate water supply and drainage outlet are needed for sinks, dish machines, steamers, **braziers** (large kettles for cooking with small amounts of liquid), waste disposals, and restrooms. In addition, adequate sewer ventilation fans should be installed to expel fumes into the atmosphere. Pipes that supply steam pressure and gas need to be sized appropriately and must meet the regulation codes of the National Plumbers Association (NPA) and the U.S. Public Service for Building and Installation.

SPACE AND KITCHEN LAYOUT

The type of service offered by a food service facility largely determines its total square footage and how it uses its space. Figure 9–1 (see page 112) shows a basic food service kitchen layout designed to use space efficiently and save workers' time and energy. Major kitchen work areas are divided into stations, or units. The number and types of different stations needed depends on a facility's volume and the types of food it serves. Supporting service areas generally include an office, rest rooms, employee lockers, and janitorial supply storage.

A large food service may have duplicate pieces of equipment and completely separate departments for different work areas. For a small food service operation, it is vital that the kitchen layout allots space for different stations as efficiently as possible. In all kitchens, the space between work stations and walk areas should be at least 4.5 to 6 feet wide. This space permits the easy transport of carts and dollies carrying food and utensils, and it prevents cross-traffic problems.

Receiving Area

The receiving area must be large enough for unloading deliveries, and it should include a space to weigh incoming supplies and to inspect for accuracy and quality. The outside door should be 40 inches to 6 feet wide to allow the passage of large cartons and equipment. To ease unloading, the floor and platform outside the receiving door should be on the same level as the floor inside the door.

Storage

Inadequate storage space can cost a food service operation time and money. The amount and use of storage space should be carefully planned to provide an appropriate amount of space plus a full view of and easy access to all stored items. The effective placement of vertical and horizontal metal shelves can ensure the full utilization of storeroom space. All paper goods and nonperishable grocery items, except chemical supplies, can be stored in a dry storage area. Food supplies must be stored at least 18 inches off the floor and there should be at least 6 inches between shelves to permit good air circulation, which helps prevent spoilage. Ventilator fans are also a necessity. Spoiled food

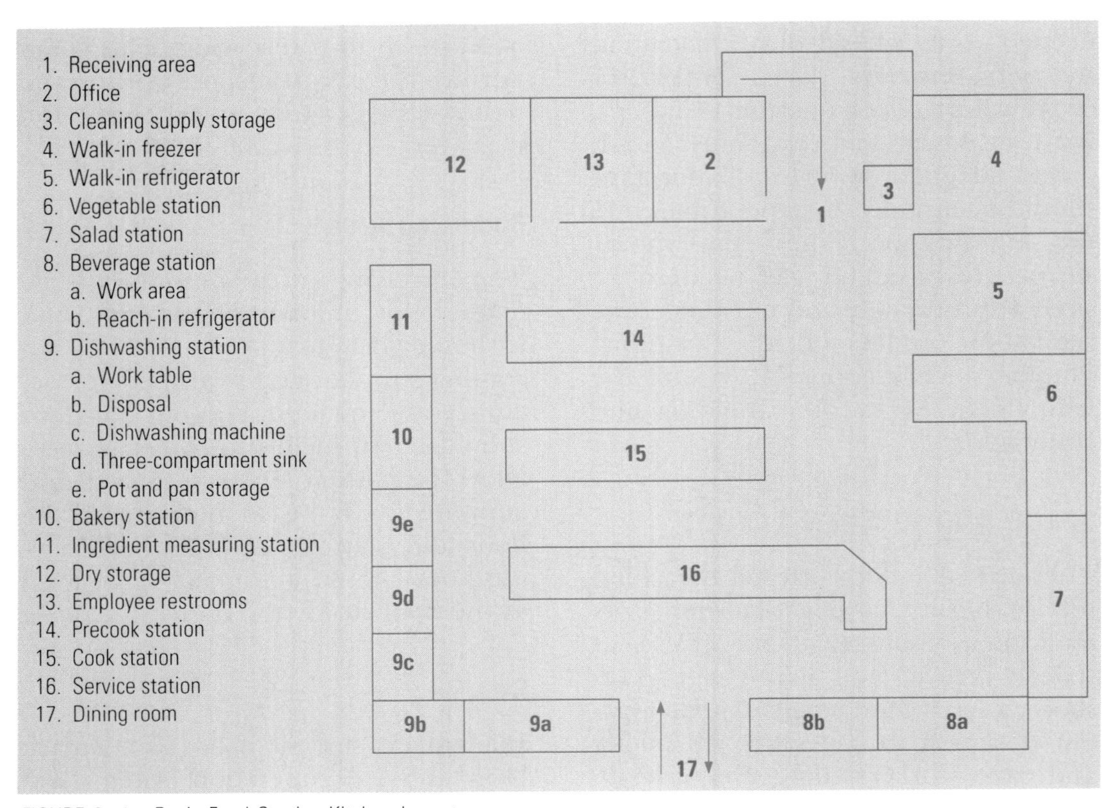

FIGURE 9–1 Basic Food Service Kitchen Layout

caused by improper storage increases the cost of served food and, in the long run, can affect a business's profits.

Walk-in refrigerator and freezer storage should be located near the receiving area, and temperatures must be monitored daily. Mobile shelving units can be used to wheel food into a walk-in for easy storage.

Vegetable Station

The vegetable station should be located close to the refrigerator. It must contain enough space for equipment, such as choppers, cutters, carts, knives, cutting boards, a two-compartment sink, work table, waste disposal, and garbage can.

Salad Station

The salad station is usually located close to the vegetable station and the serving station, which is near the dining room. This space should be large enough to include a reach-in refrigerator, a sink, knives, and cutting boards. Chopping equipment from the vegetable station can also be used at the salad station.

Main Cooking Station

The main cooking station is in the center of the kitchen and should be close to the vegetable station, the pot storage area, and the serving area. Equipment in this space includes a range, broilers, fryers, ovens, steamers, mixing machines, slicers, and a cook's table, which is usually located in front of the cooking apparatus.

A cooking station's design and equipment will depend on the facility's menu and the size of its kitchen. Small equipment, like pots, pans and utensils, can be hung from the ceiling for easy access. Large equipment may be stationary or mobile. Provisions for gas and electrical outlets should be made early in the design process to permit the use of mobile equipment. The cooking area is the main focus of a kitchen, and decisions that affect this area must be carefully considered during the early planning stages.

Baking Station

The baking station should be close to the service station. An oven, dough divider, pie roller, reach-in refrigerator, cooling racks, and a baker's table, are necessary equipment in this area. Many food service operations purchase ready-to-serve desserts. In this case, the station can be used for refrigerated and unrefrigerated display cases. Desserts are taken directly from the baking station to the service station for holding.

Dishwashing Station

The dishwashing area should be close to the dining room to save time in transporting soiled dishes. This area, however, must be built with soundproof material to avoid disturbing guests in the dining area. The dishwashing station should provide adequate space for sorting, scraping, washing, rinsing, and drying serviceware and utensils, and it should be close to the dishes and to the pot and pan storage area. The type of dishwashing equipment needed depends on the number of customers served daily and the speed at which the equipment operates. The arrangement of the dishwashing area can vary from a straight line or a square to an L shape or a U shape, and it usually flows from right to left. The dishwashing area may include a work table on which to place soiled dishes, a disposal, dishwashing machine, three-compartment sink, and pot and pan storage area. The pot and pan area storage is essential, and it should be close to the three-compartment sink and the cook station for convenience.

Serving Station

The serving station should be as close as possible to the dining area. If well designed, this area can speed service and reduce labor. The service counter should include cold and hot holding tables, a cold salad display, and space for silverware, napkins, dishes, and condiments. Beverages should be located at this station or at a nearby beverage station. The serving area must be constructed to keep hot food hot and cold food cold.

OTHER IMPORTANT FACTORS IN KITCHEN LAYOUT

In addition to planning the basic kitchen work areas, there are other factors to consider in kitchen layout. These factors include the position of kitchen equipment, adequate storage of supplies, maximum use of space, proper handling techniques, and appropriate aisle space.

Appropriate Equipment Position

Large and small equipment and utensils should be positioned around the kitchen to promote fast, easy, and efficient work. This can be accomplished by positioning a station or piece of equipment close to related stations or utensils to avoid cross-traffic problems. The strategic placement of equipment can reduce employee fatigue and increase productivity.

Adequate Supply Storage

Storing should be done according to how often items are used and how easily they can be picked up. Items used most frequently should be positioned at the front of shelves, and less-used items should be positioned at the back. Heavy items must be placed in areas where employees can remove them without injury.

Maximum Use of Space

Kitchen shelves should be long enough and wide enough to allow the easy movement of stored items, but they should not waste space. Adjustable shelving is ideal because it permits variation of spacing as needed.

Proper Handling Techniques

Good handling techniques help reduce work effort. Employees should be taught the proper way to lift heavy objects (see Figure 8-3), and loads should be limited to 35 pounds for women and 50 pounds for men. A kitchen layout must include space for mechanical devices, such as truck lifts, carts, and dollies, to lift and transport heavy objects.

Suitable Aisle Space

Clean and adequately spaced aisles are essential. Aisles should be designed with a minimum width of 4.5 feet to ensure the easy flow of work and to prevent traffic problems. Inadequate aisle space can lead to employee or customer accidents.

TYPES OF FOOD SERVICE FACILITY LAYOUTS

The kinds of food service provided today vary from simple to elegant and from affordable to expensive. The variety of food service organizations seems endless, including fine-dining restaurants, small delicatessens, coffee shops, and school cafeterias, as well as food service provided by universities, colleges, hospitals, nursing homes, airlines, and fast-food outlets. In addi-

ILLUSTRATION 9–3 The kinds of food service provided today vary from simple to elegant and from affordable to expensive. This restaurant offers a fine-dining experience.

MATH SCREEN

The number of people that can be served at one time in a food service facility's dining area is based on the number of square feet or meters required for each person seated. Space allowances vary depending on the type of facility. The following table provides the minimum space allowed per seated person for various types of food operations:

Type of Operation	Minimum Space Allowed for Seated Person	
	Square Feet	Square Meters
Table service, hotel, club, restaurant	15	1.4
Counter service	18	1.7
School cafeteria	9	.8
Commercial cafeteria	16	1.5
Banquet service	10	.9

To calculate how big a dining area must be, use the following formulas:

Square feet × number of people to be served at one time = total square feet

or

Square meters × number of people to be served at one time = total square meters

Study this example: How many square feet of space are needed to seat 75 people in a restaurant with table service?

75 people × 15 square feet per seated person = 1,125 square feet

Answer: 1,125 square feet of dining space are needed to seat 75 people at one time for table service.

Now test your math skills: How many feet of space are needed to serve 500 students in a school cafeteria?

ILLUSTRATION 9–4 The variety of food service organizations seems endless, including fine-dining restaurants, delicatessens, coffee shops, and the company cafeteria shown here.

tion to establishing goals and making sure that funds are available, a food service director must make many decisions about food service layout and design, and he or she must be familiar with the different types of service that can be provided. Some of these service types were presented in Chapter 3. Because many people work in fast-food facilities, the following section discusses the layout and design of a typical fast-food restaurant. Actual layouts will depend upon the size and type of restaurant.

NUTRITION NOTE RECOMMENDED DIETARY ALLOWANCES

The objective of the government's school lunch program is to safeguard the health and well-being of American children. Therefore, the school lunch program must meet one-third of the RDA (recommended dietary allowance) requirements for daily nutrients.

The RDAs are described in a government publication that lists appropriate nutrient intakes for people of various ages. Many plans have been developed to help the public make nutritious food choices that ensure adequate nourishment. One of the most widely used plans is the basic four food group plan. This plan includes simple guidelines that specify the types of food, as well as the quantities, that need to be consumed daily. The quantities to be consumed vary according to different age groups.

FOUR FOOD GROUPS

Group	Sample Foods	Number of Servings for School Age Children	Serving Size
Meat and meat substitute	Beef, pork, fish, eggs, nuts, legumes	2	2–3 oz cooked meat, fish, or chicken; or 1 c legumes
Dairy	Milk, cheese, yogurt, ice cream	3–4	1 c milk; or 1½ –2 oz. cheese; 1 c yogurt; or 1¾ c ice cream
Fruits and vegetables	All fruits and vegetables	4	½ c fruit, vegetable, or fruit juice
Grain	Bread, cereals, and pastas	4	1 slice bread; ½ c cooked cereal; 1 cup ready-to-eat cereal; or ½ c pasta

FAST-FOOD RESTAURANT DESIGN

In 1988, McDonald's, the number one fast-food organization, dominated all other fast-food chains with over $10 billion in gross sales. Burger King (number two in terms of sales) grossed $5 billion, and Kentucky Fried Chicken (number three) grossed $4.1 billion. The reasons for their success are many: in general, they have consistently provided better value and service to customers, made menu changes as needed, implemented more efficient operational techniques, and improved their restaurants through remodeling.

Fast-food restaurants have a limited number of menu selections, all of which are quick and easy to prepare. Menu items at different outlets include hamburgers, cheeseburgers, chicken, french fries, tacos and other popular Mexican dishes, salads, soft drinks, coffee, and milk. Most entrées are either fried or grilled. This makes ventilation an important feature of fast-food restaurant layouts.

Business volume is usually predictable, and volume purchasing is desirable because it helps reduce food cost. It is thus essential for fast-food restaurant designs to include adequate space for volume purchasing, especially for dry storage of paper goods and freezer storage of meat products.

An easy-to-follow traffic plan is also vital and should be designed with safety and speed in mind. Menu boards inside and out must be positioned where they can be seen and read easily. These menu boards should clearly describe items to prevent misunderstanding. Seating usu-

ILLUSTRATION 9–5 Burger King, the number two fast-food organization in terms of sales, grossed $5 billion in 1988.

ally includes tables with booths and tables with chairs, providing a simple but attractive decor. Garbage cans should be positioned close to the exits to make it easy for customers to dispose of unwanted food or garbage. Figure 9–2 shows a basic fast-food restaurant layout.

FIGURE 9–2 Fast-Food Restaurant Layout

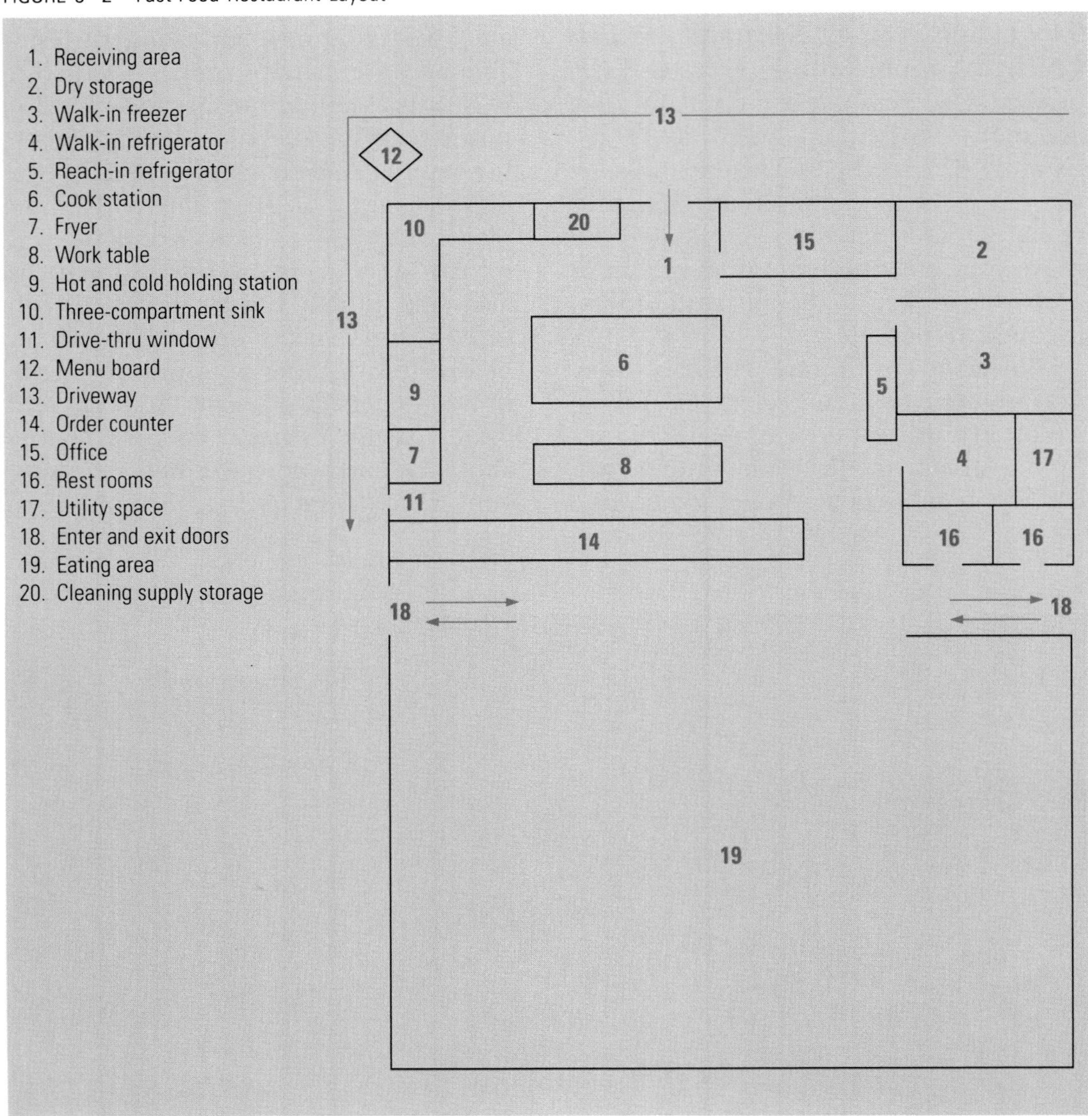

CAREER PROFILE
POSITION: DIRECTOR OF SCHOOL FOOD SERVICE

Qualifications:

- Bachelor's degree in food service management, dietetics, or related field
- Master's degree in business administration or related field desirable
- Five years experience in an institutional food service program, of which at least one year is spent in a supervisory role

Responsibilities:

- Supervises a field service manager, equipment coordinator, food service buyer, payroll assistant, commissary manager, fiscal control leader, secretary, and nutritionist
- Develops hiring, training, and evaluation procedures for all food service employees
- Analyzes new concepts in food service administration and implements new methods such as à la carte or fast-food techniques versus a traditional lunch program
- Ensures that the school district meets all the requirements of the National School Lunch Program
- Estimates the annual number of free, reduced-rate, and full-cost meals through an examination of past history, local economy, and inflation rates
- Develops budgets, prices, and labor schedules, and submits them to executive directors for approval
- Coordinates resources with local agencies to establish new and innovative programs and menus
- Serves as a member of the food service negotiation team (a team of administrators that makes important decisions with regard to purchasing products and staffing large school districts)

SUMMARY

Facility layout and design is important to the success of all food service establishments. The first thing a food service manager or owner must do is establish the organization's goals. Once the goals are clear, the manager or owner can assemble a planning team, which typically includes an architect, building contractor, engineer, and food service equipment expert. This team works on the building's design and selects materials. Important factors that must be considered during the design process include what type of walls and ceilings, lighting and wiring, and plumbing to use. The planning team also prepares a space and kitchen layout. All of a facility's work stations should be designed to complement each other during operation, allowing for safe and efficient work. The same basic layout guidelines—planning for appropriate equipment position, adequate supply storage, maximum use of space, proper handling techniques, and suitable aisle space—apply to all kitchens, regardless of the type of facility being built.

QUESTIONS

1. What are the common goals of most food service establishments?
2. What questions should be asked before designing a food service facility? List at least four.
3. A planning team of experts can assist a food service owner or director in facility layout and design. What are the specialized professions of these team members?
4. What three popular construction materials are often used in today's buildings?
5. Foods look most appealing in what type of light?
6. If natural light is not available, what two types of artificial lights are best for food service facilities?
7. With what code must all electrical wiring and installations comply?
8. All plumbing sizes and fixtures must comply with what two agencies?
9. What three types of storage are necessary in all food service facilities?
10. Food must be stored how many inches off the floor?
11. What type of shelves are best for food storage?
12. What two elements are essential in food storage areas?
13. The salad station should be near what two stations?
14. What are three areas in which the layout for a fast-food restaurant differs from other restaurants?

ACTIVITIES

Activity One Visit a restaurant in your city or town and sketch its kitchen layout. Compare your design to the design in Figure 9-1 and discuss the pros and cons of each. You and your classmates should present your designs individually on an overhead projector. You can prepare a transparency with the following directions:

- First, draw a rough draft of the design on paper.
- Use a ruler and a dark marking pen to transfer the design to a transparency.
- Neatly draw and print the different work stations within the restaurant and indicate the flow of work from the receiving area through the kitchen to the service area.

For each different layout, your class should answer the following questions:

- Are all the work areas designed to promote efficiency?
- Do the storage areas include refrigerator, freezer, dry food, and cleaning supply storage?
- What are the distances between the work areas and the storage areas?
- Are there any cross-traffic problems?
- Are the storage areas close to the receiving dock?
- How far is the service area or the dining room from the kitchen?
- Does the layout include employee and guest toilet facilities?

Activity Two Visit a fast-food restaurant and sketch its kitchen layout, following the same directions as in Activity One.

10
TOOLS AND EQUIPMENT

VOCABULARY

specification
aluminized
copperplated
range
conventional oven
convection oven
radiant energy
microwave oven
revolving oven
deep-fat frying
broilers
steam-jacketed kettle
pressure steamer
walk-in refrigerators
walk-in freezers
food mixers
food slicer

OBJECTIVES

After studying this chapter you should be able to do the following:

- Explain why written specifications for equipment are important
- Identify the large equipment used in food service kitchens
- Recognize different pieces of equipment, their special features, and the reasons for using them
- Identify the various knives used in food service kitchens and explain their different purposes
- Discuss the materials used to construct food service equipment, which materials are most popular, and why
- Outline the different types and uses of cooking utensils, food handling tools, and temperature control tools

The equipment purchased for food service kitchens should meet the standards of safety and sanitation established by the National Sanitation Foundation (NSF). Food service managers should choose equipment made of durable, easy-to-clean materials that do not give off toxic materials, odors, or colors into food. Popular materials used to construct equipment are stainless steel, aluminum, copper, iron, brass, steel, and plastic.

The first step in the purchasing of kitchen equipment is determining what equipment is needed, and then writing specifications for the needed equipment. The next section describes this process.

SPECIFICATIONS

A **specification** is a statement describing a piece of equipment requested for purchase. Food service businesses send specifications to manufacturers, who then supply equipment that meets the specifications. When planning a facility, managers or their buyers determine the specifications of the equipment they need. Before writing a specification for food service equipment, the buyer should check the state, county, and city code regulations that might influence purchase decisions. In addition, the person who writes a specification should have extensive knowledge of the use and operation of kitchen equipment.

Specifications must be written clearly, leaving no room for misunderstanding, and they should state the size, material, heat source, expected capacity, price quote, and delivery charges for the equipment. Important considerations while writing specifications include the following: (1) Expense—would a smaller or less expensive piece of equipment do the job? (2) Expansion—would this piece of equipment meet the needs for expansion? (3) Safety—can the equipment be easily cleaned and maintained with minimal work effort? (4) Profit—would purchasing this equipment help produce a profit? Well-written specifications can eliminate many operational problems by leading to equipment that saves on labor expenses, ensures maximum production capacity, and lowers the cost of maintenance.

MATERIALS USED TO CONSTRUCT EQUIPMENT

Stainless steel is the metal most often used to construct food service equipment because it resists rust and corrosion, withstands high temperatures, and is strong, shiny, and easy to maintain. It is used for all sizes of food service equipment, from large to small.

Aluminum, copper, iron, brass, and steel are also used to construct cooking equipment. These materials, however, must be treated with acid or coated with a nontoxic substance to promote food safety and easy cleaning. Steel cooking equipment is usually aluminized or copperplated to give better heat distribution because aluminum and copper are excellent heat conductors. **Aluminized** metals have been treated or coated with aluminum. **Copperplated** metals have had copper bonded to them.

Plastics have gained popularity in recent years because of their safety and versatility. Many types of plastics are used in food service, each for a different purpose. Some plastics are formed into dishwasher holders, for example, and others are molded into storage containers for hot and cold items. In addition, many types of service equipment, such as transporting units and salad bars, are now made of plastics.

LARGE EQUIPMENT USED IN COOKING

The large equipment used by the food service industry is similar to the ranges, ovens, refrigerators, and other appliances found in a typical home, only much bigger and more durable. The following sections describe these pieces of equipment.

Ranges

A **range** is designed mainly for surface cooking and the heating of food in pots and pans. Some ranges include a griddle, oven, and broiler. Ranges are generally mounted to the floor, and their framework is constructed with heavy-duty, well-braced steel. They can be gas or electric. The size of a range can vary from two to ten burners, depending on need. All ranges should be equipped with thermostatic controls that regulate surface cooking temperatures between 150°F and 550°F.

Ovens

Ovens are insulated heating units of various capacities that serve a variety of purposes. They are designed to cook food in covered or uncovered shallow baking pans, and they may contain one or several shelves depending on their size. Ovens used for roasting are approximately 15 inches high, while ovens used for baking are up to 8 inches high.

Conventional Ovens

Food is cooked in a **conventional oven** by heated air in an enclosed space. Most conventional ovens come with two shelves, each capable of holding 18-by-26-inch baking pans. This type of oven is best for baking cakes and other foods made with batter.

Convection Ovens

Convection ovens may be gas or electric and are available in different sizes. Inside a **convection oven,** a fan circulates hot air and speeds the cooking process without drying out food. This type of oven cooks food more quickly at lower temperatures. Convection ovens are popular because they are efficient and preheat in 5 minutes as opposed to 20 minutes for conventional ovens. Other desirable features of convection ovens are listed here:

- One-third less cooking time
- Lower cooking temperatures
- Increased productivity
- Less labor
- Efficient use of energy

Convection ovens, however, do create some problems, such as the following:

- Increased cooking losses
- Uneven baking (especially for cakes)
- Improperly finished meringue products
- Hard and cracked muffins

Microwave Ovens

Microwaves are an efficient source of **radiant energy,** or energy that travels in a wave motion. When a **microwave oven** is plugged into an electrical outlet, a magnetic tube in the oven converts the electricity into microwave energy, which is transmitted into the closed oven.

Microwaves pass through glass, paper, plastic, and ceramic. As microwaves bounce around the interior of an oven, they enter food from all sides and cook it all the way through at once. The more food placed in an oven, the more time it takes to cook. Another factor affecting cooking time is the food's composition. Some foods absorb energy faster than others. For example, fats and sugars absorb waves more quickly than other foods.

The key feature of microwave ovens and the reason for their popularity is the speed at which they cook and reheat food. Microwave ovens prepare food in approximately one-fourth the time of a conventional oven. They can defrost, cook, and reheat food, and they are widely used in food service kitchens, especially for reheating.

ILLUSTRATION 10–1 Microwave ovens can defrost, cook, and reheat food, and they are widely used in food service kitchens, especially for reheating.

Large Revolving Ovens

A **revolving oven** is used to bake and roast meats and bakery products in large quantities, such as for large school districts and bakeries. These ovens can hold up to seventy 18-by-26-inch pans. Shelves inside a revolving oven are built ferris-wheel style and are turned by a gear-driven mechanism outside the heating area. Each shelf is well supported and can hold five hundred pounds in weight. Panel doors on each side of the oven permit entrance to the interior for easy cleaning and repair.

Deep-Fat Fryers

A deep-fat fryer is an important piece of equipment for many food service operations. **Deep-fat frying** means cooking food in large amounts of hot fat. Many of today's deep-fat fryers are computerized and equipped with automatic pilots, thermostatic controls with a rapid recovery rate of heat, and timers that buzz when food is ready. Fryers cook food to a desired consistency and crispness without making it greasy or diminishing its flavor. Cooked food can be lifted out of the fat by mechanical devices built into the fryer. In addition, most fryers have a built-in mechanism that will clean the equipment with little effort.

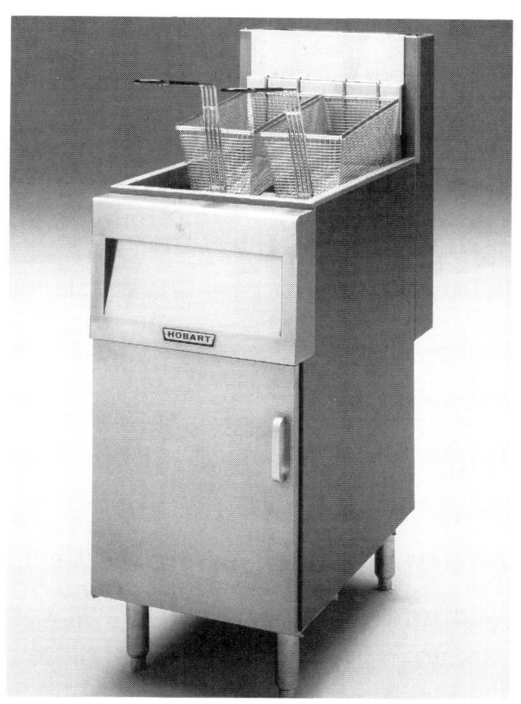

ILLUSTRATION 10–2 Deep-fat fryers cook food to a desired consistency and crispness without making it greasy or diminishing its flavor.

Broilers

Broilers cook food by exposing them to dry heat from charcoal, gas flames, or an electric metal coil. Broilers are available in different designs and sizes. The most popular design includes a waist-high grid support over the heat source.

Steam-jacketed Kettles

A **steam-jacketed kettle** uses the moist-heat method of cooking. Designed for quantity cooking, it consists of two stainless steel containers, one inside the other. Steam-jacketed kettles have an interior capacity of more than two hundred gallons. A valve inside the kettle permits steam pressure to enter the kettle and speed the cooking process. These kettles are used to cook potatoes, pasta, stews, soups, pot roast, and many other foods in large amounts. The kettle must be placed near a water supply and must have its own floor drain. Cooking with a steam-jacketed kettle requires covers, scrapers, paddles, and beaters, which are usually sold separately.

Pressure Steamers

A **pressure steamer** also cooks with moist heat, directing 5 pounds (low pressure) or 15 pounds (high pressure) of pressurized steam toward food. The steam comes in direct contact with the food, and cooks by transferring its heat and condensing water molecules derived from the food.

Pressure steamers have different capacities. They may include many compartments or just one, depending on the volume of food an establishment cooks daily. Heat from the pressure of the steam is much hotter than boiling water. Steamer covers must be opened with great care to avoid burns.

Refrigerators and Freezers

Refrigeration equipment is indispensable for food service facilities. A refrigerator is a thermally insulated cabinet or room that reduces and maintains a temperature below that of the external environment for the storage of perishable foods. A freezer maintains subfreezing temperatures for the rapid freezing and storing of perishable foods.

Several types of refrigeration units are generally used in food service. **Walk-in refrigerators** and **walk-in freezers** are the size of small closets or entire rooms. In addition to walk-in units, there are reach-in units, display and counter refrigerators, refrigerated mobile units, shake and yogurt machines, and ice machines available in various styles.

Food Mixers

Because of their versatility, **food mixers** are frequently used pieces of equipment in food service kitchens. These machines come with many attachments, each performing a different function. Mixers can mix, beat, knead, whip, slice, strain, chop, grind, and even heat and cool mixtures. They can turn liquids into solids and solids into liquids.

Mixers can sit horizontally or vertically. Vertical mixers are the most common type in the United States. They are available in different sizes, ranging from 5 to 140 quarts, and may be mounted on a wall, countertop, or floor.

Mixers with more than a 20-quart capacity are mounted on the floor and come with variable speeds, mixing bowls of different sizes, and a variety of agitation accessories. The accessories used most often include flat beaters, wire whips, dough hooks, and pastry knives.

Food Slicers

The purpose of a **food slicer** is to cut or slice foods to a consistent thickness. Slicers come in a range of sizes. Larger models may be motor driven; smaller models are operated manually. The cutting apparatus is generally a carbon steel or cast stainless steel disk from 10 to 18 inches in diameter, with an attached sharpener that keeps the blade sharp.

Slicers must have a safety guard to protect workers from accidentally turning on the machine. Two or more speeds are available on most slicers. These speed controls can be regulated

ILLUSTRATION 10–3 The purpose of a food slicer is to cut or slice foods to a consistent thickness.

to produce from one slice per minute to over fifty slices per minute. Food slicers are used on boneless meat products, like roast beef, turkey, and ham, as well as on cheeses and vegetables.

SMALL EQUIPMENT AND TOOLS USED IN COOKING

In addition to large equipment, a variety of cooking utensils, hand tools, measuring tools, temperature control instruments, and storage containers are essential for efficient food preparation and service. Other small pieces of equipment found in commercial kitchens include coffeemakers, toasters, and beverage dispensers.

Utensils

Cooking utensils are needed for range-top and oven cooking. Some of the most frequently used cooking utensils are described in Figure 10–1 (page 128). After food is cooked, it often must be strained and drained. Figure 10–2 (page 129) shows utensils used for these procedures.

Hand Tools

Hand-operated kitchen tools play a part in virtually every kitchen task. They are used when slicing, paring, sectioning, turning, beating, mashing, sifting, scraping, boning, cutting, splitting, tossing, folding, brushing, sharpening, and performing a number of other jobs.

Knives

Knives are the most important tools in a food service kitchen. Today, good knives are made of stainless steel because this metal alloy is durable and stays sharp for a long time. All professional chefs have their own set of knives, and they are usually quite offended if someone borrows them.

A good set of knives can improve work quality and efficiency. Knives must be kept in good working condition. Food service operators send their knives out to be serviced on a weekly or biweekly basis, depending on how often they are used. A butcher's steel can be used to sharpen edges between scheduled maintenance times.

Knives are classified according to their uses. Each knife is designed for a specific purpose. A set of knives should include those shown in Figure 10–3 (see pages 130–31).

a. Stockpots
These large, 2.5 to 40-gallon pots with short, sturdy handles and high walls are used to make soups and stocks. The larger pots are equipped with a faucet for easy pouring. Stockpots are used to cook a variety of sauces, soups, and stews on top of a range.

b. Saucepots
Saucepots are large, heavy-duty pots. They come in a variety of sizes from 6 to 60 quarts and are used to cook sauces, stews, and soups on top of a range.

c. Double Boilers
These utensils are two containers in one. The bottom container is deep and holds hot water, and the top container, in which food is cooked, is shallow and is held suspended over the hot water by the rim of the bottom container. Double boilers are used to prepare cream fillings and puddings.

d. Braziers
Braziers are large, round, wide, shallow pots used for braising, browning, and stewing meats.

e. Roasting Pans
Roasting pans are rectangular, large, and shallow. They come in different sizes and depths. The most commonly used commercial size is 18 by 26 inches.

f. Frying or Sauté Pans
These pans may be from 7 to 16 inches in diameter. They have shallow, sloped walls and are used to sauté meats and vegetables.

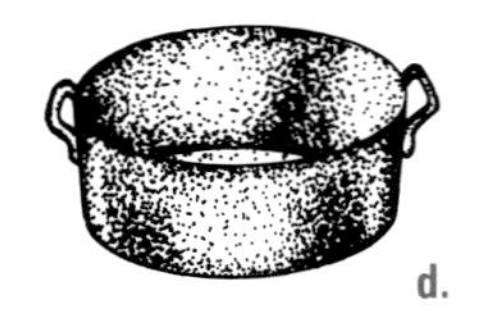

g. Saucepans
Saucepans resemble saucepots but are smaller and have one long handle. They come in different sizes and are used to prepare creams and sauces.

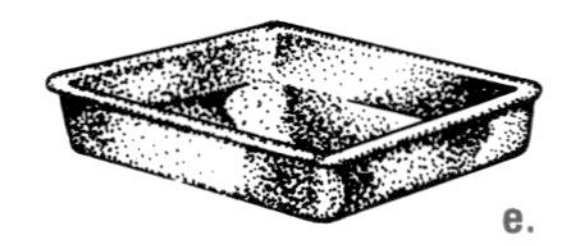

FIGURE 10–1 Cooking Utensils

a. China Caps
These cone-shaped stainless steel bowls are used with wooden mallets, which force food through mesh cones. They are used for making purees and straining gravies and sauces.

b. Colanders
Colanders are perforated bowls used to rinse fruits and vegetables and to strain spaghetti and other pastas.

c. Sifters
Sifters are used to sift flour and other dry ingredients used in baking.

a.

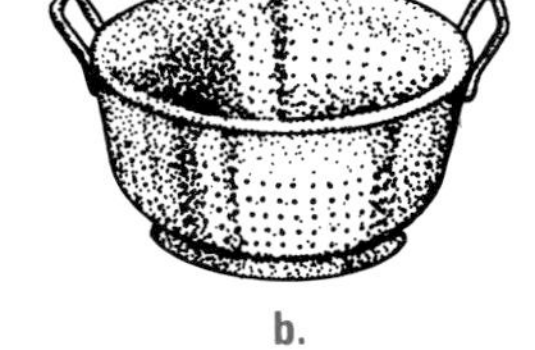

FIGURE 10–2 Straining and Draining Utensils

Food-handling Tools

Food is handled in a number of different ways during the preparation stages. Figure 10–4 (see pages 132–33) provides information about the various tools used to handle food.

Measuring Tools

Measuring cups are available in various sizes and in metal, glass, and plastic styles. On most cups, graduations are imprinted in red for easy reading and include both U.S. and metric standards. Clear plastic cups are especially nice because they are durable, dishwasher safe, stain and odor resistant, and they allow the user to see the contents. Measuring cups come in two ba-

ILLUSTRATION 10–4 Measuring cups are available in various sizes and in metal, glass, and plastic styles. A variety of glass measuring cups is shown here.

sic designs; one design includes a lip, which makes it easy to measure liquids, and the other, which eliminates the lip, is used to measure dry ingredients.

Temperature Control Tools

In order to avoid overcooking or undercooking food, temperature control tools must be used. Several different styles of thermometers are illustrated in Figure 10–5 (see page 134).

Storage Containers

Storage containers are important for retaining the freshness and quality of various foods. They also keep pests and rodents out of the food supply, thus cutting down on waste and the possibility of food contamination.

FIGURE 10–3 Knives

a. French or Chef Knife
This knife is used for chopping, dicing, slicing, mincing, julienning, and shredding raw fruits and vegetables. It is used more than any other knife.

b. Carving Knife
A carving knife is used to slice roast beef, ham, and many other meats. Cutting can be done in a vertical or a horizontal motion. The knife must be held straight while pressure is applied to permit even slicing.

c. Boning Knife
This knife is used to remove bones from meat. Deboning is done in short and quick backward strokes. This knife is also used to disjoint poultry and dice raw meats.

d. Paring Knife
A paring knife is used to trim the skin from fruits and vegetables. When paring, the food is held in one hand and the knife is held in the other.

e. Bread Knife
A bread knife has a serrated edge and is used to slice bread and cakes. Food is held down with one hand while the other hand uses the knife in a light, sawing motion.

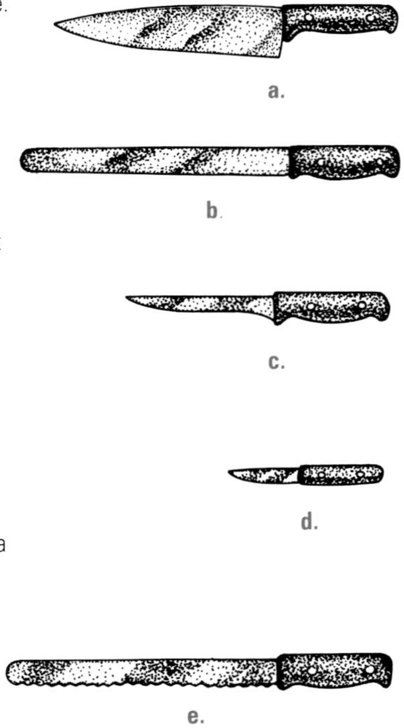

a.

b.

c.

d.

e.

f. Butcher Knife
A butcher knife is used to cut large pieces of raw meat into sections.

g. Fruit and Salad Knife
This knife is slightly larger than a paring knife and is used to cut fruits and vegetables into sections.

h. Cleaver
A cleaver is used to cut through the bones and joints of meats, poultry, finfish, and shellfish.

i. Butcher Steel
A butcher steel is used to sharpen the blade edges of knives.

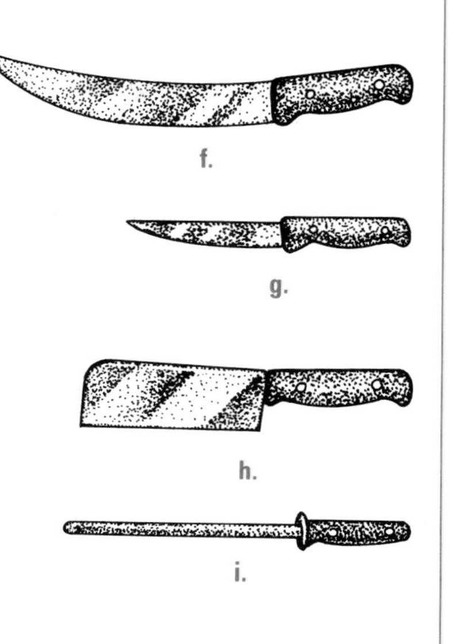

FIGURE 10–3 (continued)

MATH SCREEN

Determining equivalent amounts of various food measurements is an essential cooking skill. The following chart lists equivalent amounts for many common measurements.

1 tablespoon	=	3 teaspoons	=	0.5 ounce
2 tablespoons	=	6 teaspoons	=	1 ounce
4 tablespoons	=	¼ cup	=	2 ounces
5⅓ tablespoons	=	⅓ cup	=	2.8 ounces
8 tablespoons	=	½ cup	=	4 ounces
11 tablespoons	=	⅔ cup	=	5.4 ounces

16 tablespoons	=	1 cup	=	8 ounces
32 tablespoons	=	2 cups	=	1 pint
1 quart	=	4 cups	=	2 pints
½ gallon	=	8 cups	=	2 quarts
1 gallon	=	16 cups	=	4 quarts

Study this example to see how to determine measuring equivalents: How many teaspoons are in 1 cup?

1 cup = 16 tablespoons

1 tablespoon = 3 teaspoons

Answer: $16 \times 3 = 48$ teaspoons in 1 cup

Now test your math skills: How many pints are in 1 gallon?

a. Meat Tenderizer
A meat tenderizer is made of aluminum or wood. Its pounding surface has teeth that help tenderize tough meat fibers.

b. Grater
A grater is used for shredding cheeses, the rinds of fruits, and other foods.

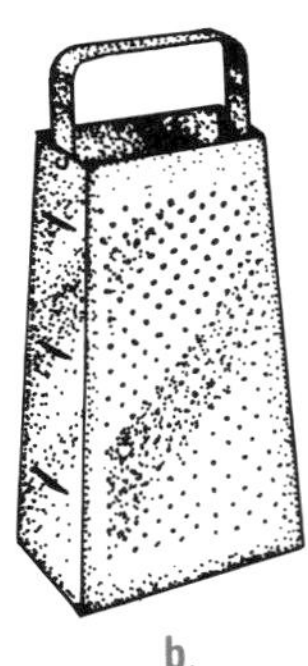

c. Dough Cutter or Scraper
This tool is used to cut dough into sections and for scraping baking tables and meat blocks clean.

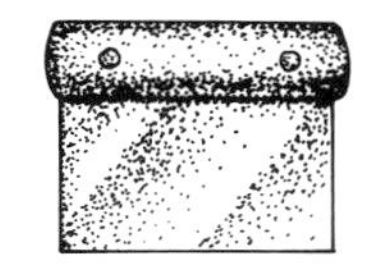

d. Peeler
A peeler has a blunt blade that peels off only the skin of vegetables and fruits.

e. Egg Slicer
An egg slicer is used to slice hard-boiled eggs. A peeled egg is placed inside the slicer and sharp wires press against the egg to cut.

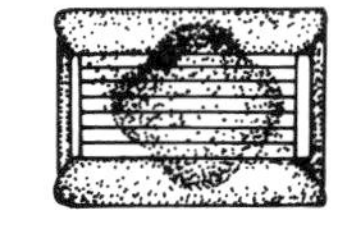

f. Melon Ball Cutter
This small metal scoop is used to cut round shapes of butter, fruits, vegetables, and semisolid sauces.

g. Pastry Brush
A pastry brush is used to brush butter and egg mixtures on baking products.

h. Wire Whip
This tool is made with heavy wire and is used for mixing heavy sauces and gravies.

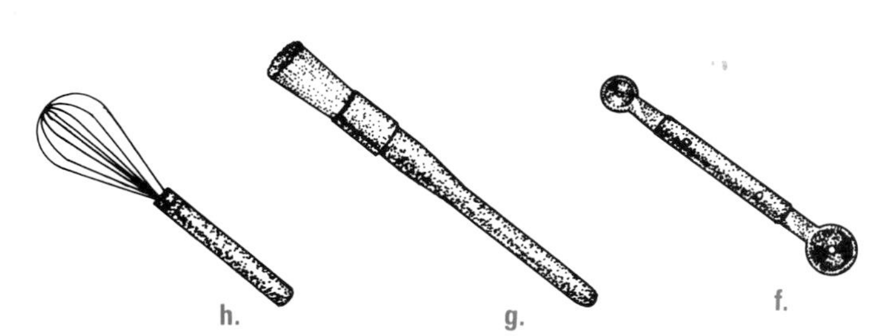

FIGURE 10–4 Food-handling Tools

i. French Whip
This whip is made with lightweight wire and is used for lightly whipping egg whites and creams.

i.

j. Large Spoons
Large spoons are used to mix or serve food. A slotted spoon is used to drain liquids or solids.

k. Ladle
A ladle is used to serve soups, sauces, and gravies. Ladles come in sizes that hold from 2 to 8 ounces.

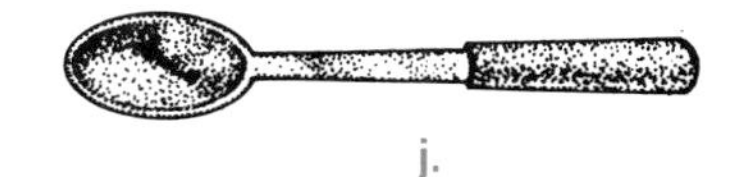

l. Skimmer
A skimmer has a slotted design in the center and is used to remove unwanted solids from soups.

m. Cook's Fork
This tool is used to turn or remove large pieces of meat from pans and to hold meat firmly in place while slicing.

n. Turner
A turner is used to turn hamburgers, pancakes, grilled-cheese sandwiches, and omelets. It may be slotted or solid.

o. Measuring Spoons
These tools are used to measure ingredients with accuracy. Measuring spoons come in five sizes: 1/8 teaspoon, 1/4 teaspoon, 1/2 teaspoon, 1 teaspoon, and 1 tablespoon.

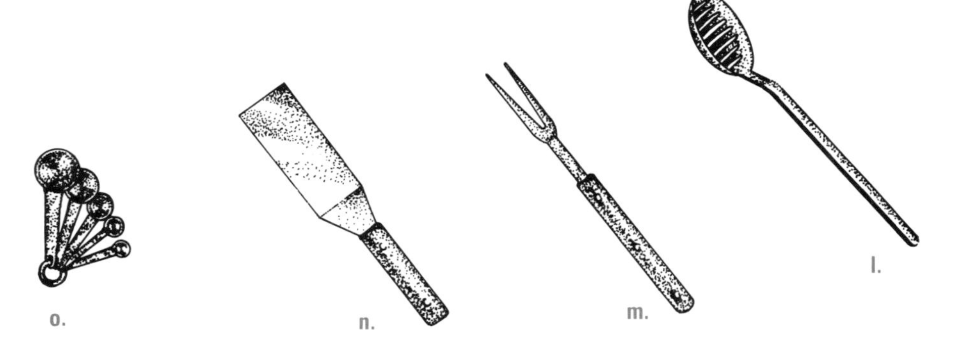

Figure 10-4 (continued)

a. **Confectionery or Candy Thermometer**
This type of thermometer is used to make candies and frosting.

b. **Probe Thermometer**
A probe thermometer is used to measure the internal temperature of meats.

c. **Oven Thermometer**
This thermometer is used to measure oven temperatures, which often are not the same throughout an oven and may need to be adjusted.

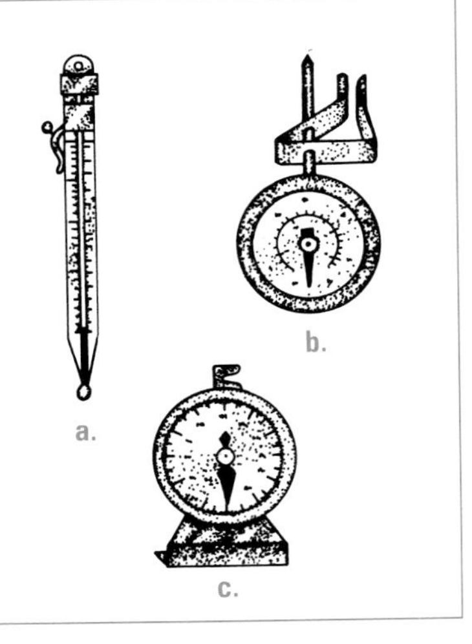

FIGURE 10–5 Temperature Control Tools

Ingredient Bins

Ingredient bins are usually made of durable plastic with heavy-duty casters, which make them easy to transport. They fit neatly under work tables and are easy to clean. Ingredient bins are used to store large quantities of flour, sugar, lettuce, dry legumes, and many other items.

Dish Carts

Dish carts provide a compact, sanitary place to store dishes and flatware. Heavy-duty casters make them easy to maneuver, and strong plastic construction ensures that they are durable, safe, and easy to maintain.

Food Storage Boxes

Food storage boxes come in different sizes and are made of hard plastic materials that can withstand temperatures from −40°F to 212°F. They are dishwasher safe and have a clear finish that permits food identification. These boxes can be used for wet and dry storage.

ILLUSTRATION 10–5 Ingredient bins are used to store large quantities of flour, sugar, lettuce, dry legumes, and many other items.

NUTRITION NOTE VEGETABLE STORAGE

One serving of green, leafy vegetables, such as spinach, mustard, collard, or turnip greens, provides generous amounts of vitamins A and C. In addition, greens contain iron and calcium. All of these vitamins and minerals promote good health. Proper storage of green, leafy vegetables ensures optimum quality and nutrient retention.

Green vegetables maintain their crispy freshness and most of their vitamin and mineral content if they are stored in covered containers or plastic bags and placed in a refrigerator. Sealed containers protect other foods from picking up the greens' strong odor. All refrigerated greens must be used as soon as possible to assure nutrient retention.

When storing large quantities of frozen greens, the freezer temperatures must be below 0°F. Greens may be kept frozen up to eight months. Unopened canned vegetables must be stored in a cool, dry place and should be used within one year from the date they were purchased to ensure quality and nutrient retention. After opening, canned vegetables should be used within one to two days.

ILLUSTRATION 10–6 Food storage boxes come in different sizes and are made of hard plastic materials that can withstand temperatures from −40°F to 212°F.

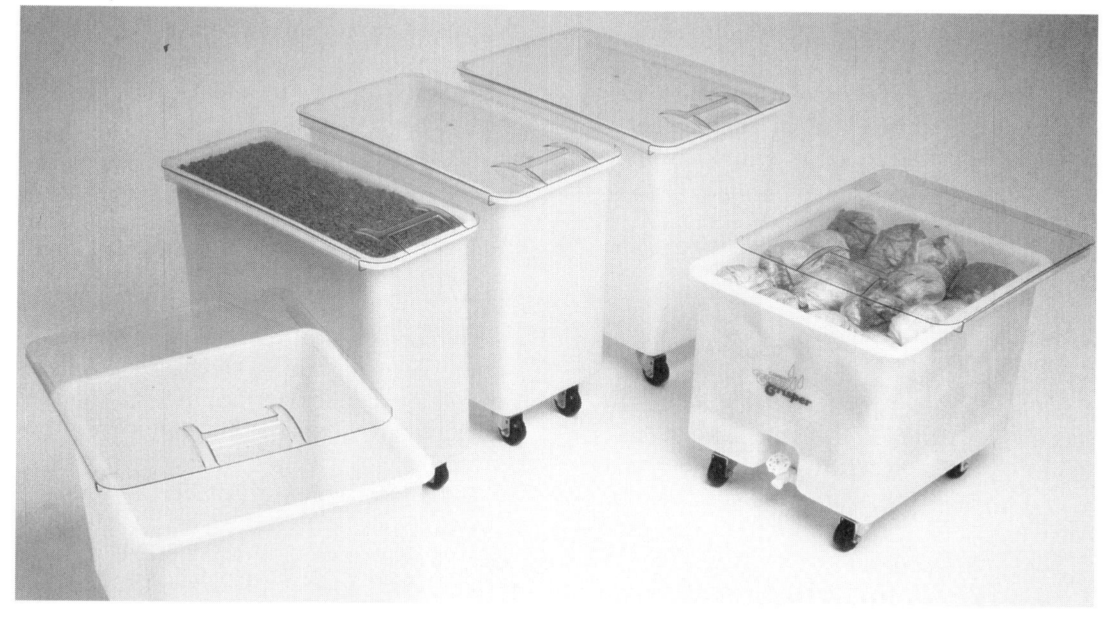

CAREER PROFILE
POSITION: SCHOOL FOOD SERVICE EQUIPMENT COORDINATOR

Qualifications:

- Good verbal and written communication skills
- Two years of experience in the area of purchasing equipment and supplies or an equivalent combination of training, education, or experience that meets the minimum qualifications

Responsibilities:

- Reviews and prioritizes equipment repair requests
- Prepares work orders and arranges for the repair of kitchen equipment
- Conducts inventories of small and large equipment at sites
- Reviews orders for small equipment and distributes kitchen utensils to schools
- Informs vendors of the bid process and the criteria they must meet to be considered as potential vendors
- Prepares, reviews, and revises bid specifications for needed food service equipment and utensils
- Verifies that vendors meet bid specifications and prepares comparable-cost analyses on vendors and items
- Reviews comparable analyses and makes recommendations to supervisor
- Forecasts future needs for equipment and utensils
- Arranges for the installation of new or replacement equipment
- Prepares purchase orders for new equipment
- Makes recommendations to the director on the placement and type of equipment in new kitchen designs
- Prepares comparable-cost analyses on the repair of equipment versus the purchase of new equipment and submits figures to the food service director
- Performs related duties as required

SUMMARY

Food service equipment is expensive. Before purchasing equipment, food service operators must carefully assess their facility's goals and then prepare specifications for equipment that is tailored to the needs of their establishment.

Because of the safety concerns surrounding certain metals, the materials used to construct equipment is an important factor when purchasing kitchen tools. Other considerations include expense, production capacity, labor needs, maintenance, future expansion, and profit potential.

Large equipment must be chosen carefully to avoid costly mistakes. A refrigerator that does not chill food correctly or a range that is too small to meet a facility's needs can seriously affect an operation's profits. Small tools are equally important;

good utensils, hand tools, and storage containers can encourage efficient and creative work.

Food service employees should be trained in the proper use and maintenance of kitchen equipment. All tools must be handled with care and a concern for safety. The manufacturer is the best source of information on the employment and maintenance of equipment. When selling new equipment, most manufacturers will send out an equipment specialist, on request, to demonstrate operational and cleaning procedures. The success of a food service facility depends to a large extent on its tools and equipment and how well they are taken care of.

QUESTIONS

1. What is a specification, why is it important, and what should be included in a specification for equipment?
2. What are two materials commonly used to construct food service equipment and why are they popular?
3. Why are some pieces of cooking equipment coated with a nontoxic substance? Why are other pieces aluminized or copperplated?
4. What are the main purposes of a range and of an oven?
5. How does a convection oven differ from a conventional oven?
6. What types of materials are used for microwave cooking?
7. What type of heat method is used when one cooks with a broiler and what are some of the heat sources used?

Question 8 includes three types of knives and descriptions of four knives. Place the letter of the correct definition in the blank to the left of each type of knife.

8. _____ paring knife
   _____ chef knife
   _____ cleaver
   a. used to cut through the bones and joints of meats, poultry, fish, and shellfish
   b. used to slice meat products
   c. used to chop, slice, and dice
   d. used to trim off the skin from fruits and vegetables
9. How does a wire whip differ from a French whip?
10. What are the uses of probe thermometers and oven thermometers?

ACTIVITIES

Activity One Study one piece of equipment found in your classroom kitchen or a restaurant in your city or town. Examine the equipment and explain or demonstrate to your class the following:

- The name of the equipment
- The materials from which the equipment is constructed
- What the equipment is used for
- How the equipment is used

Activity Two Display kitchen hand tools, including knives, on a table or counter and mark each one with a different number. If your classroom does not have all of the tools, use pictures from this chapter. Provide each of your fellow students with a handout that names all the tools. Ask them to identify each tool by placing in the appropriate space the number of the actual or pictured tool.

PART FIVE

MENU PLANNING

11

MENU DEVELOPMENT

VOCABULARY

menu
courses
continental breakfast
à la carte
table d'hôte
brunch
main course
aperitifs
cordials
portion size
menu copy
menu layout
typeface
merchandising copy
accent copy
descriptive copy

OBJECTIVES

After studying this chapter you should be able to do the following:

- Explain the importance of the menu to the success of a food service operation
- Outline the process of selecting menu items
- Detail the contemporary course structure for both formal and informal restaurant menus
- Discuss the different types of menus and the components of menu design and layout
- Give examples of three types of menu copy

The principle product of any food service operation is food. Restaurants, unlike many businesses, do not display their products for the customer to review and discuss before purchasing. Because of the perishable nature of the product, refrigeration or heat is required for both attractive appearance and food sanitation. The only way most restaurants have of telling a customer about their product is through their **menu.** A menu is a printed form that outlines all of a restaurant's offerings in a series of **courses.** Courses are groups of food items that are served in an established sequence during a meal.

As discussed in Chapter 1, meal service has changed dramatically since the time of the ancient Greeks and Romans. By the 1800s, the French had established the menu format that is used today. This format breaks down a meal into a seven-course sequence. In Figure 1–2, a menu outline developed by Auguste Escoffier shows each of the seven courses, from appetizer to dessert. This menu format is used for special banquet functions and also in extremely formal restaurants. Most modern commercial menus, however, involve far fewer courses and are more abbreviated. The following section discusses the menu formats used today. These formats are derived from the classical French format.

MENU FORMAT

Commercial menus employ many variations on formal menu formats for breakfast, lunch, and dinner. Depending on the style of a food service operation, the management can offer a number of menu format combinations, including banquet and appetizer menus. The following sections discuss these different menu formats.

Breakfast

Breakfast can be offered in a variety of combinations. Most restaurants provide individual portions of each item as well as groupings of items ranging from a **continental breakfast,** which consists of fruit or fruit juice, breakfast bread, and a beverage, to a complete eight-course meal. A full eight-course breakfast menu includes food presented in the following sequence: juice, fruit, cereal, eggs, meat or fish, vegetables or starch, bread, and beverages. An example of such a meal would be orange juice, sliced honeydew melon, bran flakes, scrambled eggs, bacon, home fried potatoes, rye toast, and coffee or milk. A variation involving only five courses would include grapefruit juice, waffles with sliced strawberries, sausage links, breakfast rolls, and coffee.

Restaurants very often incorporate breakfast combinations into their menu formats as well as providing individual menu items. Those menu items that are offered separately are priced **à la carte,** or individually. The combination breakfasts are served **table d'hôte,** or all together for one price. Chapter 12 discusses menu and food pricing in more detail.

Luncheon

American luncheon menus are offered in both formal and informal formats. Historically, luncheon has often included as many courses as dinner.

In some countries, luncheon is still the main meal of the day. The modern American life-style, however, has eliminated the full luncheon format except for special occasions and in formal restaurants.

The full-course or formal luncheon menu structure contains eight courses. These courses are as follows: an appetizer; a soup; a salad; eggs; fish, poultry, or meat; a vegetable or starch; a dessert; and a beverage.

American businesses generally give their employees a lunch break of 45 to 60 minutes, allowing only enough time for a quick meal. As a result, a number of quick meals have evolved. One example is the sandwich, which combines a number of ingredients representing the courses in a formal luncheon format. A hamburger, for instance, is usually prepared in a six-ounce to an eight-ounce portion and is served in a roll with sliced tomato, lettuce, and, often, melted cheese. The plate presentation generally includes a form of fried potato, usually french fries. The meal thus incorporates a salad, meat, and vegetable into one course. Beverages are served at the same time, and an optional dessert completes the two-course meal.

Salads and soups have become popular luncheon items. They can be served quickly and accomodate diet- and health-conscious patrons. Salad bars and buffets are common variations of the luncheon format. **Brunch** is the combination of breakfast and luncheon items, and it usually includes fruit, eggs, fish or meat, breads, dessert, and beverages.

Dinner

Dinner has become the most formal meal for American diners. It retains more of the formal course structure than either breakfast or luncheon. A nine-course dinner format is found on most full-service restaurant menus. The courses consist of an appetizer, soup, salad, fish, poultry, meat, vegetable or starch, dessert, and beverage.

American dinner menus offer a variety of ways to abbreviate this course structure. Most restaurant patrons, for example, use the appetizer course as a complement to the main course, choosing a fish, poultry, or beef item according to their entrée selection. Soup is often used as an appetizer.

Salad is the third course. Salad was originally intended to cleanse the palate or freshen the taste buds after the heavy tastes of sauces, meats, and fish served as part of the entrée course. Today, salads are more elaborate. They start with a base of the wide range of greens available in modern produce markets and add vegetables, fruits, nuts, herbs, and a variety of other foods. Salads are sometimes featured as entrée items, with meat, poultry, or fish as a major ingredient.

Entrées, or the fourth course, can be served with vegetables and starches in a plate presentation. In the formal menu tradition, a dinner may include a series of entrées: a fish course fol-

Cold Appetizers

Shrimp Cocktail 6.95
Littlenecks on the Half Shell 6.25
North Atlantic Seafood Medley 6.95
Artichoke Hearts Tosca 5.25
Sesame Beef 4.95

Hot Appetizers

New England Quahog Pie 4.95
Stuffed Mushrooms Mont D'Or 5.95
Coconut Shrimp Caribe 6.95
Herbed Ribbon Noodles with Escargots 6.25
Petite Wellington Béarnaise 6.95

Salads

Caesar Salad 3.95
Audrey's Dinner Salad 1.95
Fisherman's Melange 4.95

Soups

A selection of two homemade soups served with sea toast. 2.25

Entrées

Rhode Island "Mini Bake" 18.95
☆ Grilled Block Island Swordfish 16.95
☆ Sole Monterey 14.95
☆ Filet of Beef Nicoise 18.95
Western Style Sirloin Steak 17.95
Veal Oscar 16.95
Veal Saltimbocca 15.95
Duckling with Summer Fruits 15.95
☆ Chicken Parisienne with Assorted Mushrooms 13.95
Shrimp Stuffed Shells 12.95

☆ *These items may be prepared according to your dietary needs.*
The above entrées are garnished with the Chef's own creations.

Cigarette smoking only is permitted in designated areas.

UNIQUELY AUDREY'S

As a progressive restaurant utilizing only the finest of fresh seasonal ingredients, Audrey's food preparation is unique in many aspects. Fresh pasta, breads, rolls, pastries, ice cream and sherbet are created on the premises. Our chef offers a weekly selection of creative dishes which includes a lite entree using no salt, cream or fat for our diet-conscious patrons.

Audrey's menu reflects our uncompromising commitment to excellence in providing for your dining pleasure.

ILLUSTRATION 11–1 Dinner has become the most formal meal for American diners.

lowed by a poultry course and then by a meat course. Public dining habits today have changed this tradition by using the appetizer for one food item, such as fish, and the **main course,** or entrée, for another item, such as poultry or meat. Certain restaurants, however, continue to offer a formal nine-course menu.

Although listed as a separate course, vegetables and starches are generally served with the entrée. A few cuisines treat pasta, rice, or grains as single courses and serve them separately.

Dessert, the eighth course, is one of the most profitable areas of a menu. Desserts have become increasingly popular with the American public in the past few years. Items in the "sweet section" often include pies and cakes, classical pastries, souffles, mousse, ice cream, fruits, and cheeses.

Most menus offer a wide variety of beverages, such as juices, soda pops, grain alcohols, coffees, and teas. Beverages served at the beginning of a dinner are called cocktails, or **aperitifs.** When nonalcoholic drinks are desired, juices or soda pop can be provided.

Wine may be offered before and during the meal. After-dinner drinks, or **cordials,** are served following the meal to aid digestion. Coffee and tea are offered at the end of the meal.

Banquet Menus

Banquet menus provide the same meal to a group of people being served in a private function room. Each person receives the same foods for each course. Banquet menus can be adapted to serve from ten to five hundred or more guests.

Appetizer Menus

Another way to vary the formal dinner menu is to offer appetizer-sized portions of many different foods. Guests can order three, four, or five items, which may or may not include soup and salad, in place of an entrée. Wines are offered by the glass to provide a wide range of taste combinations.

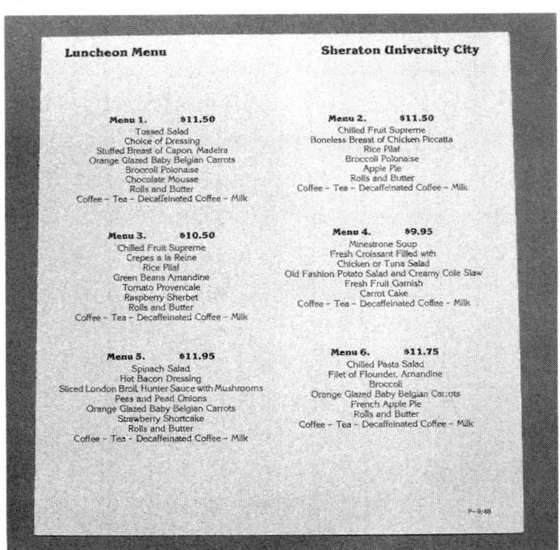

Luncheon Menu

Sheraton University City

Menu 1. $11.50
Tossed Salad
Choice of Dressing
Stuffed Breast of Capon, Madeira
Orange Glazed Baby Belgian Carrots
Broccoli Polonaise
Chocolate Mousse
Rolls and Butter
Coffee - Tea - Decaffeinated Coffee - Milk

Menu 2. $11.50
Chilled Fruit Supreme
Boneless Breast of Chicken Piccatta
Rice Pilaf
Broccoli Polonaise
Apple Pie
Rolls and Butter
Coffee - Tea - Decaffeinated Coffee - Milk

Menu 3. $10.50
Chilled Fruit Supreme
Crepes a la Reine
Rice Pilaf
Green Beans Amandine
Tomato Provencale
Raspberry Sherbet
Rolls and Butter
Coffee - Tea - Decaffeinated Coffee - Milk

Menu 4. $9.95
Minestrone Soup
Fresh Croissant Filled with
Chicken or Tuna Salad
Old Fashion Potato Salad and Creamy Cole Slaw
Fresh Fruit Garnish
Carrot Cake
Coffee - Tea - Decaffeinated Coffee - Milk

Menu 5. $11.95
Spinach Salad
Hot Bacon Dressing
Sliced London Broil, Hunter Sauce with Mushrooms
Pees and Pearl Onions
Orange Glazed Baby Belgian Carrots
Strawberry Shortcake
Rolls and Butter
Coffee - Tea - Decaffeinated Coffee - Milk

Menu 6. $11.75
Chilled Pasta Salad
Filet of Flounder, Amandine
Broccoli
Orange Glazed Baby Belgian Carrots
French Apple Pie
Rolls and Butter
Coffee - Tea - Decaffeinated Coffee - Milk

P-9/88

ILLUSTRATION 11–2 Banquet menus provide the same meal to a group of people being served in a private function room.

MENU CONTENT

Menu content refers to the selection of food items in each menu category. A well-balanced menu and established portion controls are crucial to successful food production and service.

Menu content is determined by a five-stage process. This process results not only in a final printed menu but also in a group of alternate items that can be used as specials. Figure 11–1 illustrates the menu item selection process.

In the first step of the process, all the items that could be included in a particular menu category are listed. During the second step, items on the list that could pose purchasing, preparation, or service problems are eliminated. In the third step, the list is reduced to those items suitable to the restaurant's cuisine or theme. During the fourth step, items within an established price range and that the restaurant can produce well are chosen. In the fifth step, the final selections are determined for each category on the printed menu.

Portion Control

Planning menu content involves not only developing lists of items but determining the general portion size for each category of menu item. The **portion size** is the amount of a food item served per order. To establish pricing guidelines, it is necessary to develop a range of portion sizes. For example,

NUTRITION NOTE CYCLE MENUS

Institutions such as hospitals, schools, and retirement and nursing homes have different requirements for menu planning than do commercial restaurants. Many corporations provide cafeterias and dining rooms for employees. The customers in these situations usually make use of the same food service on a daily basis. In order to vary their menus without serving different meals every day, many institutions develop cycle menus. These menus enable institutional food service departments to provide nutritionally balanced meals at a lower cost. Cycle menus rotate meals over a set period of time, such as 10 days, 12 days, or 21 days. To establish a cycle menu, a set of main items that generally receive a favorable response from customers is selected. These items are then scheduled to appear in a staggered pattern so that they are not served on a predictable basis.

cuts of beef for entrées can be in portions of 6, 8, or 12 ounces. The surrounding items, however, such as potatoes and vegetables, will be the same portion size regardless of the size of the beef. The price is adjusted only according to the beef portion.

Salads are priced according to plate size. One luncheon plate presentation may involve only 6 ounces of greens. Another, filling a 12-ounce salad bowl, will have a higher food cost and selling price. Appetizers such as shrimp and escargot are priced according to the number of pieces in a portion. Stews, chowders, and soups are priced by the cup or bowl. Cakes and pies are portioned by the slice. Ice cream costs are determined by the size of the scoop, and pudding prices are

FIGURE 11–1 Menu Item Selection Process

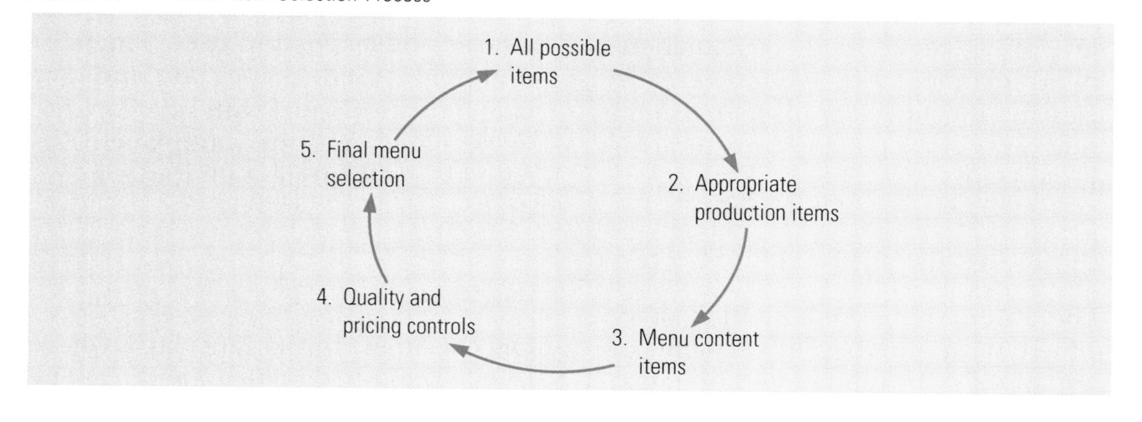

ILLUSTRATION 11–3 To establish pricing guidelines, it is necessary to develop a range of portion sizes. Pudding prices are calculated by dish size.

calculated by dish size. Beverages are priced by the standard glass sizes of 6, 8, and 12 ounces as well as by the amount and type of alcohol, if applicable.

Creating Menu Items

The challenge of creating cost-effective, easy-to-produce menu items is an exciting part of menu planning. In addition to researching cookbooks and developing new food combinations, menu planners can explore creative uses for the many preprepared items now offered by the food manufacturing industry.

MENU DESIGN AND COPY

The overall design of a menu is determined by a number of elements. A restaurant's central theme (often organized around entertainment or a cuisine), its style of service, and its price ranges can all influence the design of a menu. **Menu copy** is the text that is used to introduce and describe the menu items to the customer. Menu copy that accurately and creatively describes items can increase sales by raising customer interest in different courses and foods.

Menu Layout

The way in which illustrations, graphics, and the meal course structure are arranged on a menu is called the **menu layout.** This visual presentation of the menu items can determine what choices a customer makes. Items that are easy to locate or attractively designed on the menu are often selected before other items.

Menus can be laid out in a number of formats. Figure 11–2 shows an example of a well-designed menu.

This sample menu is organized so that a customer can locate sections of the menu easily, and it highlights the entrée specials. Menu layouts come in a variety of shapes and sizes. The most commonly used shapes are the two panel, the two panel with side flap, the single panel, and the multipanel. Figure 11–3 illustrates all these menu shapes.

Illustrations

Once a menu format and layout have been established, illustrations can be

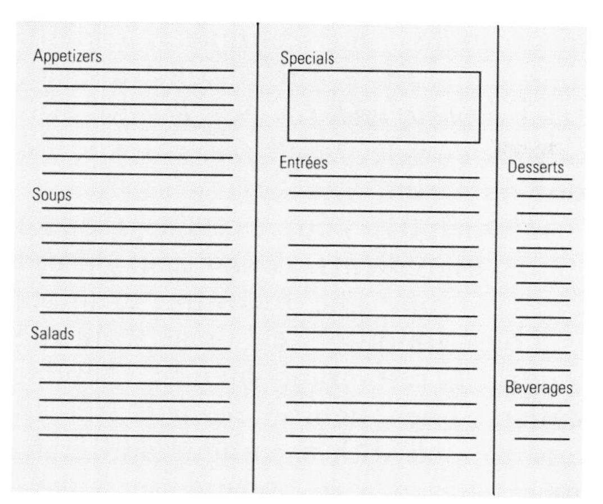

FIGURE 11–2 Sample Menu Layout

added. Menu illustrations should create interest and reinforce a restaurant's theme and cooking style.

Typeface

Typeface is the style of printing type that is used in a menu layout or other printed piece. Three different areas of a menu require careful consideration when selecting a typeface: course headings, item names, and item descriptions.

Course headings can be printed in a decorative typeface to complement a menu's overall design. Ideally, this typeface should direct a customer's eye to key sections of a menu. Item names should appear in a typeface that can be easily read but that also acts as a decorative element. Item descriptions must be set in a readable typeface to help a customer in item selection. In general, the best typeface styles are the simplest ones.

Menu Copy

The goal of menu copy is to create customer interest and reinforce servers by supplying information about item content. There are three types of

FIGURE 11–3 Menu Shapes

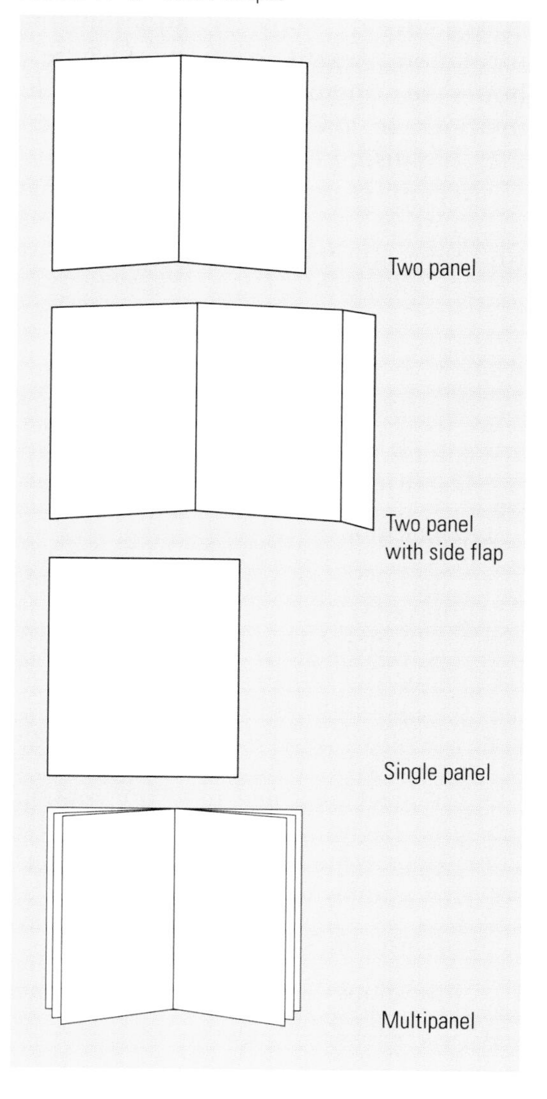

copy on most menus: merchandising copy, accent copy, and descriptive copy.

Merchandising copy is information that creates interest in the restaurant. Historical information, profiles of the owners, or a discussion of the featured cuisine are popular types of merchandising copy. In addition, information about dining facilities, hours of operation, and acceptable methods of payment are considered merchandising copy.

Accent copy creatively highlights menu items. Accent copy is often used to draw attention to the names of menu items. One example is a menu item name for a stuffed tomato salad, *I Say Tomato and You Say Tomato.* Another example, *Dog Gone,* is for a hot dog on a bun.

Descriptive copy explains a menu item's ingredients, cooking style, and presentation. The following example is a description of veal oscar from the Audrey's restaurant menu: "Tender cutlets of veal lightly breaded, sautéed and topped with asparagus and crabmeat, glazed with hollandaise sauce." Note how the copy includes many descriptive words to appeal to a customer's senses.

CAREER PROFILE
POSITION: MENU PLANNER

Qualifications:

- Bachelor's or associate's degree preferred
- Nutrition knowledge required
- Education in food service and food preparation desired
- A variety of restaurant experiences required
- Management and food preparation experience necessary
- Good writing skills and a knowledge of graphic design essential

Responsibilities:

- Develops a repertoire of menu items that can be prepared by the food production team
- Prepares recipe and cost cards for each menu item with the help of the executive chef
- Establishes a menu pricing format
- Selects a menu layout appropriate to the establishment's course structure and operational style
- Writes menu copy to accurately describe menu items and create customer interest
- Measures the success of the menu by analyzing the sales statistics, food cost percentages, and customer comments

SUMMARY

One of the most critical factors in restaurant success, a menu must be carefully planned, organized, and developed. A menu's look and style helps to establish the tone of a restaurant.

Menu format, which depends on a facility's method of service, tells customers the number and types of courses available. The type of menu items that an establishment offers determines menu content. Menu items are selected through a five-step process. Portion control also affects menu content, and is primary to successfully pricing menu items as well as to balancing portion sizes between the appetizer and the entrée courses.

Menu design has an important influence on customers. How well a menu is designed determines how effectively it sells food items. The major elements of menu design are layout, illustrations, and typeface.

Menu copy is used to describe the restaurant and its menu items to customers. The purpose of menu copy is to create interest in the restaurant and the food items being offered as well as to accurately describe the ingredients and preparation styles used for each menu item. The three types of menu copy are merchandising copy, accent copy, and descriptive copy.

QUESTIONS

1. Why is a menu's quality important to the success of a food service operation?
2. What items would be on the menu of a typical continental breakfast?
3. What steps are involved in the menu selection process?
4. What is the difference between *à la carte* and *table d'hôte* service?
5. What is the structure of a full-service dinner?
6. Define *portion control.* What does portion control help to determine about items on the menu?
7. What are the major elements of menu design?
8. What are the four most common menu shapes?
9. What are the three sections of a menu that use typeface? Define each area.
10. What are the three types of menu copy?

ACTIVITIES

Activity One Using the course structures illustrated in this chapter, develop a breakfast, luncheon, and dinner menu.

Activity Two Choose a restaurant theme based on either a cuisine, a mode of entertainment, or a degree of formality. Using the menu selection process, develop menu items for each meal and course.

Activity Three Select a menu shape and lay out a course outline. If you have done Activity Two, choose menu items for the layout from the results of that activity.

12

RECIPE STANDARDIZATION AND FOOD COSTING

VOCABULARY

recipe
yield
portion
food costing
recipe card
file number
cost card
food cost percentage (FC%)
cost of purchased unit (CPU)
extension
cents factor (CF)
portion cost (PC)
abstract
selling price (SP)
profit
overhead
labor
food cost

OBJECTIVES

After studying this chapter you should be able to do the following:

- Discuss the importance of using recipe cards and cost cards in all types of food service operations
- Write a recipe card, detailing ingredients, specifying weights and measures, and stating cooking methods
- Define food costing and the various terms associated with this process
- Outline a cost card, detailing the four major sections of the card
- Identify the three basic costing equations and the three costing factors involved with each one

A **recipe** is a set of written instructions for making a food item from specific ingredients. A recipe is similar to a chemistry formula. When a recipe is followed exactly, a consistent product results. If a recipe's measurement, mixing, temperature, and cooking directions are heeded, it should **yield,** or produce, a specified number of portions. A **portion** is the amount of the finished product of a recipe that is served to a customer. It is the cost per portion that is used to calculate the selling price of a menu item.

Food costing is the process of determining the cost of a recipe's ingredients. A recipe's total food cost is the basis of a number of formulas used to establish a profitable selling price. In addition to food cost, the selling price of a menu item depends on labor, overhead, and desired profit.

Accurate recipe cards and cost cards are the backbone of successful restaurant management. As discussed in Chapter 11, choosing a good balance of menu items is essential to developing an attractive and profitable menu. If these menu items, however, are not produced in the kitchen exactly the same way every time, then both food cost and customer satisfaction will suffer. It is essential for good restaurant management that any employee who helps produce a food item be familiar with that item's recipe card and cost card.

FOOD COST CONTROL

The use of accurately developed recipe and cost cards helps food service establishments control food costs—an important aspect of profitability. Strict control of food costs can contribute significantly to the financial success of a food service business.

Recipe Cards

A **recipe card** is a detailed outline of the ingredients and directions needed to produce a food item. It should accurately and clearly state the name and the weight or measure of each ingredient as well as the necessary steps for successfully preparing the item. A recipe card must also indicate how many portions the recipe will produce. It also serves as the basis for the information on a cost card. Usually typed on heavy paper index cards, recipe cards can now be stored in a computer by specifically designed software programs, reviewed on the computer screen, and printed out when copies are needed. Figure 12–1 (see page 152) shows an example of a completed recipe card.

Recipe cards are broken down into three major sections: the recipe title and other information in the top section, the column headings, and the body of the card. These sections are described in the following paragraphs.

Recipe Title

The recipe title should be printed clearly in capital letters. If the title of a recipe is in a foreign language, then the English translation is placed in parenthesis underneath the title.

File Number

The top section of a recipe card also includes the card's file number. The **file number** refers to the number a recipe has been given in an organiza-

Recipe title: CHICKEN CURRY (Serve with rice creole #R-5)		File #MC-P4	Yield: 20 portions Portion: 2 pc/6 oz sauce
Ingredients	**Weight**	**Measure**	**Method**
Chicken, fryers cut up (cut breasts in half) Butter	12 lb 7 oz		1. Sauté chicken pieces in butter
Salt Onion, chopped Apples, sliced	 1 lb 4 oz 	2 tbsp 8 small	2. Add chopped onion, sliced apple, and sauté
Curry powder Chicken stock		4 tbsp 3 1/2 pt	3. Dust mixture with curry powder 4. Add chicken stock to cover 5. Cook 30-40 minutes until tender 6. Remove chicken, hold stock aside
Cornstarch Cream Lemon juice		3 tbsp 2 1/2 c 4 tbsp	7. Add cornstarch, cream to thicken 8. Add lemon juice 9. Reduce sauce until thickened

FIGURE 12–1 Completed Recipe Card

tion's total file system of recipe cards. File numbers are assigned first by menu category and then by number. In the entrée, or main course, category, items are listed by main ingredient, such as fish, poultry, beef, or pork. For example, roast turkey would be listed as MP–07 (Main course/Poultry–seventh item in this category). File numbers correspond to a complete list of available recipes by arranged menu category in file or on a computer.

Yield and Portion

Information on yield and portion appears in the top section of a recipe card. Yield refers to the number of individual portions that a recipe makes. Portion is the amount of a recipe used for each individual serving. For example, if a recipe is for a pie, and if the yield is 8 slices, then each portion is one-eighth of the pie. In a recipe for 6.25 quarts (200 ounces) of chicken vegetable stew that serves 25, the yield is 25 portions and each portion is 8 ounces.

Column headings and the card body, the other two sections of a recipe card, are located under the top section containing the recipe title, file number, and yield and portion. The column headings and body divide the card into the four components of each recipe: ingredients, weight, measure, and method.

Ingredients

Ingredients are the food items used in a recipe. In Figure 12–1, for example, chicken (in the form of cut-up fryers, or young chickens) is the first ingredient. The items needed to sauté the chicken are then listed. Vegetables are the next ingredients, including a short description of how they should be prepared (the onions are chopped and the apples are sliced). Curry and chicken stock are the next ingredients to be used, combined with cornstarch, cream, and lemon juice.

Weight

Weight is the amount of an ingredient that should be used to produce the specified number of portions noted in the top section of a recipe card. Ingredients are placed on a scale to determine weight. Liquids as well as solids can be weighed. It is important, however, to calculate the weight of the container in which a liquid ingredient is weighed.

Measure

Measure refers to the amount of any ingredient that is not weighed on a scale. Measure can be specified by the size or number of cans, bunches, pieces, or other items. Both dry and liquid ingredients can be measured by the cup, tablespoon, and teaspoon.

Method

Method is the prescribed way that a product is to be put together. The method column details directions for preparing, combining, and cooking ingredients. Before a recipe card is written, the chef or manager should work through the recipe and outline each step. Directions should be written as clearly as possible, using the simplest vocabulary terms. Each step should be matched with the corresponding ingredients and their weights and measures to help the kitchen staff to accurately produce the recipe. Numbering each step and separating some of them with lines makes it easier for cooks to keep track of where they are in the production of a recipe.

Cost Cards

A **cost card** is an established form that outlines the cost of each ingredient in a recipe. When all of these costs are totaled, the cost of each portion of a recipe can be determined. To establish a profitable selling price for each menu item, a food service manager must make sure that the item's cost to the business represents an appropriate percentage of the selling price. That percentage is called the **food cost percentage (FC%).**

Cost cards, like recipe cards, are usually typed on heavy paper index cards, but they can also be stored in a computer and printed out when copies are needed. Figure 12–2 (page 154) shows a completed cost card.

A cost card is broken down into four major sections: the recipe title

Recipe title: CHICKEN CURRY		File #MC-P4	Yield: 20 portions Portion: 2 pc/6 oz sauce	
Ingredients	Weight	Measure	CPU	Extension
Chicken, fryers	12 lb		1.19 lb	14.28
Butter	7 oz		2.35 lb	1.05
Salt		2 tbsp	.24 lb	.02
Onion	1 lb 4 oz	8 small	.40 lb	.52
Apples		3 1/2 pt	.79 lb	3.16
Chicken stock			.96 qt	1.68
Curry powder	2 oz		.64 oz	1.28
Cornstarch		3 tbsp	.69 lb	.07
Cream		2 1/2 c	2.10 qt	1.40
Lemon juice		4 tbsp	1.92 qt	.12
PC: $1.20 Abstract: 2%=.02 Total PC: 1.22		FC%: 30% SP: $4.07	Extension total: 23.58 CF: 2%=.47 Total: $24.05	

FIGURE 12–2 Completed Cost Card

and other information in the top section, the column headings, the body of the card, and the calculation box. As on a recipe card, if the title of the recipe is in a foreign language, then the English translation is placed in parenthesis underneath the title. The file number, as well as the information on yield and portion, should correspond to the recipe card on the same item.

Likewise, the ingredient, weight, and measure columns on a cost card relate to the same item's recipe card. The ingredient column should list the ingredients in the same order as they are listed on the recipe card. The weights and measures may have to be converted if the purchase unit of a food product is different from the unit of measurement specified on the recipe card. This conversion ensures that the prices listed in the other two columns on a cost card, the CPU and the extension columns, will be accurate. The following paragraphs discuss these two columns in detail.

CPU

The **cost of purchased unit (CPU)** is the price that an organization pays for a unit of food. A unit is the measure or weight that an ingredient comes in at the time of purchase. Table 12–1 shows some of the different units in which ingredients can be purchased.

Items that are purchased by the package or the case are usually broken down for costing by the piece or by

INGREDIENT	PURCHASE UNIT
Tomatoes, crushed	#10 cans
Green pepper	Pound
Olive oil	Gallon
Basil	Bunch
Oregano	Package

TABLE 12–1 Purchase Units

the pound. Eggs are priced by the purchase unit of a dozen. The price of a dozen eggs is calculated by dividing the total cost of the case by the number of dozens in the case. The cost of one egg can then be determined by dividing the cost of one dozen eggs by twelve. This calculation is shown in Figure 12–3.

Extension

Extension is the cost of the amount of an ingredient used in a recipe. This cost is calculated by multiplying or breaking down the CPU according to the ingredient's weight or measure. Often, the units used in the weight and measure columns and the CPU column are not the same. These columns must be converted to the same units before the extension cost of an ingredient can be found.

The numbers found in the extension column are used to calculate the figures found in the bottom section, or calculation box, of a cost card. These figures include the extension total, the cents factor, the portion cost, the abstract, and the food cost percentage, which are used to determine the menu item's final selling price. The following paragraphs discuss these different figures in detail.

FIGURE 12–3 Egg Cost Calculation

Purchase unit: Case (12 dozen eggs)
Case cost: $9.48
Cost of a dozen: $9.48 ÷ 12 = $0.79
Cost of an egg: $0.79 ÷ 12 = $0.0658 or 7 cents per egg

Extension Total

After each ingredient has been costed by weight or measure the extension column is totaled. Onto this total is added a **cents factor (CF).** The CF is a percentage of the extension total that represents any small costs for ingredients that are not included on the cost card, such as a pinch of spice or a tablespoon of flavoring. The percentage used to calculate a CF is determined by management and is usually 1 or 2 percent of the extension total. A CF is always added, regardless of the recipe ingredients, to ensure that all food costs are covered. An item's cost may increase slightly or a cook may use more of an item than specified in a recipe. Adding a CF helps to cover these unexpected costs.

Portion Cost

The **portion cost (PC)** is calculated by dividing the extension total (plus the CF) by the number of portions yielded by a recipe. The PC for the recipe in Figure 12–2 is calculated as follows:

Extension total (plus CF): $24.05
Number of portions: 20
$24.05 ÷ 20 = $1.20

Abstract

An **abstract** is the cost of any food or seasoning not included in the recipe but used in the service of an item. Salt, pepper, condiments, sugar, extra breads, and crackers are typical items included in an abstract. Two percent of the total portion cost is the average figure used for an abstract. If management feels, however, that costs for these items are higher, then the percentage is increased accordingly. In fast-food restaurants, the abstract often includes all paper and plastic goods, such as napkins, straws, stirrers, and utensils, as well as condiment packets. The abstract percentage for fast-food items can be as high as 5 percent. The PC and the abstract are added together for the total PC.

Food Cost Percentage

Food cost percentage, or FC%, is the percentage, determined by management, that the cost of a food item should represent of its selling price. For example, if the selling price of a menu item is $10 and the food cost percentage is 30 percent, then the food cost for the item should not be more than $3. Methods of calculating food cost and determining the FC% are covered later in this chapter.

ILLUSTRATION 12–1 In fast-food restaurants, the abstract often includes all paper and plastic goods, such as napkins, straws, stirrers, and utensils, as well as condiment packets.

Selling Price

Selling price (SP) is the price at which an item is sold. The selling price should cover all costs and ensure a profit. If a food cost percentage has been determined and the food or portion cost is known, then a selling price can be calculated by dividing the PC by the FC%. Figure 12–4 shows an example of this calculation.

Once a selling price is determined, it is evaluated according to a number of factors. These factors are discussed in the next section.

Cost cards must be constantly reviewed and updated to ensure that fluctuating purchase prices are reflected in the selling prices. Although menu prices are usually not changed on a daily basis, the portion size or presentation of an item can be altered to help reduce the overall food cost if, for example, one ingredient is increasing in price. Using a computer with a food service software program makes accurate costing much easier. A computer can automatically adjust the food costs on cost cards, based on daily invoices and purchase prices.

Total portion cost (PC): $1.82
Food cost percentage (FC%): 30%
PC ÷ FC% = SP
$1.82 ÷ 30% = $6.07

FIGURE 12–4 Calculating Selling Price

This updated information enables management to be immediately aware of food and menu items that could cause cost problems.

FOOD COST MATH

Food cost math, or the calculation of a food service operation's food costs and menu prices, involves three steps. The first step is to determine food cost percentage: what percentage an item's cost should be of its selling price. The second step is to calculate the cost of the food. The third step is to establish a selling price based on the information from steps one and two. In some cases, selling price is determined before the cost of food is established. There are a number of mathematical equations that can be used during any of these three steps. Which equation a food service manager chooses depends on the information that he or she has concerning step one, two, or three.

The following section will explain the most commonly used methods in food cost math. To successfully perform food cost math, it is important to understand all the terms and their abbreviations introduced in this chapter's discussion of recipe and cost cards.

Food Cost Math Calculations

To establish how much a food's cost should represent of its sales price, four factors involved with producing the item must be known. These four cost factors are food, labor, overhead, and profit. Each of these factors is assigned a percentage of the final selling price. Figure 12–5 (page 158) shows the breakdown of $1 of food sales into these four categories to illustrate the percentages of cost generally used throughout the food service industry. **Profit** is any money left after all expenses are paid, including taxes. **Overhead** refers to the operating expenses of a business, such as rent, taxes, utilities, and insurance. **Labor** is the direct cost of employing help,

ILLUSTRATION 12–2 To establish how much a food's cost should represent of its sales price, four factors involved with producing the item must be known. These four cost factors are food, labor, overhead, and profit.

Profit:	$0.10	10%
Overhead:	$0.20	20%
Labor:	$0.30*	30%
Food cost:	$0.40*	40%
Total sale:	$1.00	100%

*These two costs fluctuate depending on the local cost of labor and the cuisine of the establishment. Industry experts recommend that the total of these costs not exceed 70 percent.

FIGURE 12–5 Breakdown of a Food Sale

meaning hourly wages and salaries. It also includes the indirect costs of employee benefits, insurance, Social Security, and training. **Food cost** is the cost of any food or food-related item used in an operation, including nonalcoholic beverages. The food cost for a particular dish is the total portion cost (the portion cost plus the abstract) as calculated on a cost card.

Three simple equations to determine cost, selling price, or food cost percentage can be used if two of the three factors are known. Figure 12–6 outlines these three equations.

The final step in food cost math is establishing a menu price. The calculated selling price will not always be a good menu price. It is important that a customer feels that a menu price represents a value that they perceive is equal to the menu item. This means that a selling price may have to be adjusted either up or down, depending on the food cost and what management feels is a customer's perceived value for the item. The menu price may also vary slightly from the selling price on a cost card if it needs market adjustment.

Menu Marketing

Menu marketing strategy does not always rely on a profit margin for each item to determine selling prices. An 8-ounce or 12-ounce cut of prime rib of beef, for example, may be priced below the desired food cost percent-

MATH SCREEN

In order to determine if they have reached their cost goals, managers must calculate the total of the costs of food, overhead, and labor. This figure is then subtracted from the total receipts or sales for a given time period to determine if a profit has been made. In the following problem, calculate the actual percentages that costs represent of sales and determine if there is a profit. Refer to Figure 12–5, which outlines the ideal percentages that each factor should represent of sales.

Overhead:	$ 270.00	%
Labor:	$ 405.00	%
Food cost:	$ 540.00	%
Profit:	$ 135.00	%
Total sale:	$1,350.00	100%

TO FIND	YOU MUST KNOW	MATH FUNCTION
Food cost (FC)	Selling price and food cost percentage	SP x FC% = FC
Selling price (SP)	Food cost and food cost percentage	FC ÷ FC% = SP
Food cost percentage (FC%)	Food cost and selling price	FC ÷ SP = FC%

Sample problems:

The selling price of a menu item is $9.50 and the food cost percentage should be 35 percent. How much money is available for food cost?

$9.50 x .35 = $3.32

The food cost listed on a cost card is $3.32 and the desired food cost percentage is 35 percent. What is this item's selling price?

$3.32 ÷ .35 = $9.49

The selling price of a menu item is $5.00 and the food cost listed on its cost card is $2.75. What is the food cost percentage for this item?

$2.75 ÷ $5.00 = .55 or 55 percent

FIGURE 12–6 Three Costing Equations

age as a special incentive to bring customers into a restaurant. If enough of these items are sold, it may actually reduce the price that the restaurant pays for the beef, thus maintaining the original food cost percentage. The prime rib could also be featured with a variety of promotions for beverages or à la carte menu items that have higher selling prices and lower food costs, which would offset the loss on the beef price. Menus are reviewed periodically in order to find out which items are not selling. These items can either be removed or adjusted to suit customers' needs and desires. (More information about food service marketing is discussed in Chapter 23.)

Menu Pricing

Menu pricing is based on the theory that food service customers do not select menu items in the same way that they choose retail items such as clothing or supermarket food. Menu prices are rounded up or down to the half- and quarter-dollar mark, instead of being set at odd amounts. Menu prices ending in *95 cents,* such as $7.95,

ILLUSTRATION 12–3 Menu pricing is based on the theory that food service customers do not select menu items in the same way that they choose retail items such as clothing or supermarket food.

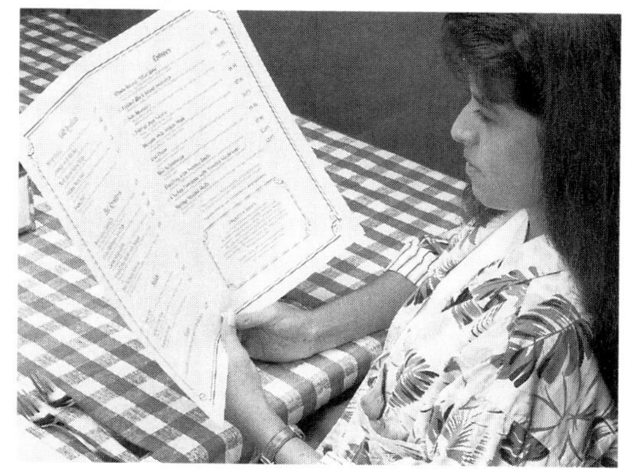

$9.95, or $11.95, are also acceptable.

Menu pricing is part of a restaurant's overall marketing program. Each menu price should be reviewed in relation to its actual selling price, to the customers' perceived value for the item, to what the competition is charging for the same item, and to how important the item is to the menu's overall success.

CAREER PROFILE
JOHN S. WALTER, CCE, CPC: RESEARCH CHEF, PRODUCT DEVELOPMENT AND TECHNOLOGY, CAMPBELL SOUP COMPANY

John Walter's interest in food service began when as a high school student he served as a bus person and held other kitchen positions. Mr. Walter then attended the Culinary Institute of America's (CIA) chef training program. After this training, he participated in a chef's apprenticeship program sponsored by the American Culinary Federation (ACF), earning ACF certifications as a Culinary Educator (CCE) and Pastry Chef (CPC). He later received a bachelor's degree in food service management from Johnson & Wales University.

Throughout his culinary training, Mr. Walter pursued an interest in baking and pastry. He became skilled in decorative pastry arts and pulling sugar. In 1980 he joined the staff of the International Baking and Pastry Program in the Culinary Division of Johnson & Wales University, and later became assistant director.

Currently Mr. Walter is with the Campbell Soup Company as one of a small group of chefs who specialize in research and development. His principle responsibility is to develop signature menu items, such as soups, entrées, and side dishes, for the company's national food service accounts.

SUMMARY

The profit margin in food service businesses is generally based on volume, and it is reasonably small if calculated per menu item. It is thus extremely important to monitor food costs and selling prices. Recipe cards and cost cards are effective management tools for this purpose, and they must be reviewed and updated daily. If these cards are not developed properly, a food service manager will be unable to analyze recipes and costs in a timely and effective manner.

To use and develop recipe and cost

cards, food service managers must be familiar with terms and abbreviations such as FC%, CPU, CF, extension, and abstract. It is also important for managers to understand weights and measures and how to convert one to the other. In addition, the ability to calculate food cost, food cost percentages, and selling prices is essential to food service management. Computer programs can now complete these calculations and update information as food and operating costs change. Because food products are highly perishable, the cost of wasted food can easily erase any profits.

Food, overhead, and labor costs determine management's ability to make a profit. A knowledge of how much these different factors should represent in an item's selling price is necessary for successful food service.

QUESTIONS

1. Why are recipe cards important to successful kitchen production and management?
2. Why are cost cards a key to profitable food service management?
3. What is the primary difference between weight and measure?
4. In what ways can measure be stated?
5. Why is a recipe card like a chemical formula?
6. How do food service managers calculate the cost of a portion of food?
7. State and define the four major factors involved with producing a menu item.
8. Define *food cost.*
9. There are three equations used for food costing. What is the unknown factor, and which equation should be used to find it, in each of the following statements? Solve the equation for each.
    a. A menu item's selling price is $10.00 and its food cost percentage is 35 percent.
    b. The food cost for a menu item is $2.50 and the desired food cost percentage is 30 percent.
    c. The selling price of an item is established as $8.00 and its food cost is $3.00.
10. How should final menu prices be established?

ACTIVITIES

Activity One With the help of your teacher, select a recipe for which to write a recipe card. Using Figure 12–1 as a guideline, detail the ingredients and specify weights, measures, and methods.

Activity Two Using a menu from a local restaurant, determine what food cost percentages the management used to determine menu prices. Select items from each section of the menu for your calculations.

Activity Three Outline a cost card, detailing the four major sections of the card. If you completed Activity One, transfer the ingredients, weights, and measures onto the cost card. With the help of your teacher, determine the CPU from current supermarket prices. Calculate the extension and do the remainder of the cost card calculations. Determine the selling price based on a 35 percent food cost percentage.

13

PURCHASING, RECEIVING, AND STORING

VOCABULARY

purchasing
forecasting
United States Department of Agriculture (USDA)
United States Department of Health and Human Services (USDHHS)
Food and Drug Administration (FDA)
Federal Trade Commission (FTC)
Fair Packaging and Labeling Act of 1967
fixed bid
weekly bid
contract
specification
purchase requisition
purchase order
invoice
perpetual inventory
physical inventory

OBJECTIVES

After studying this chapter you should be able to do the following:

- Describe why it is important for a buyer to comply with state and federal regulations
- Explain the different procedures used in purchasing by bid and by other methods
- Identify the different sources of information used to assist in vendor selection
- Identify the types of record forms used in purchasing and explain the differences between them
- Explain the appropriate procedures for receiving and storing food and supplies

Purchasing, receiving, and storing foods and supplies are essential components of high-quality food production and service. An establishment's buyer must be knowledgeable, aware of the current market, and experienced in her or his food service area to be able to forecast future changes that may affect the business. In addition, a buyer should understand federal and state regulations that apply to food quality, production, processing, labeling, transportation, and safety. Knowledge of different purchasing methods and appropriate receiving and storing procedures can help managers ensure the success of their businesses.

PURCHASING

Purchasing is the act of acquiring goods for a given price. Food service management personnel responsible for purchasing must be knowledgeable and up-to-date on the current products available, and they must be able to forecast future needs based on the market and economic conditions. **Forecasting** is the ability to estimate the what, where, how, and how much of future product purchasing, based on previous experience and current knowledge. In addition, managers must be aware of the federal and state regulations concerning the safe production, processing, and transportation of food products.

ILLUSTRATION 13–1 Food service management personnel responsible for purchasing must be knowledgeable and up-to-date on the current products available, and they must be able to forecast future needs based on the market and economic conditions.

Federal and State Regulations

The food service industry is tightly controlled by state and federal regulations. The majority of these regulations are enforced by the **United States Department of Agriculture (USDA)** and the **United States Department of Health and Human Services (USDHHS).** In recent years, several acts were passed by Congress to give these agencies the power to ensure food safety, quality, and accurate food labeling.

The **Food and Drug Administration (FDA)** is a government agency responsible for enforcing the Food, Drug, and Cosmetic Act of 1938. The act, which has been amended several times, establishes several regulations covering the production, manufacturing, and distribution of foods that are involved in interstate commerce (except for meat and poultry, which are regulated by the USDA). The FDA checks whether or not food labels are accurate and ensures that manufacturers include required information on their labels.

NUTRITION NOTE NUTRITION LABELING

When purchasing food products based on nutritional content, it is essential to read and compare label information. Decisions on what items to purchase depend on the type of diet for which the food is prepared. For example, people with high blood pressure need to decrease the sodium content in their food. People with a weight problem need to decrease the amount of fat and total calories. Whatever the reason for concern, reading a product's nutrition label can help a buyer make wise purchasing decisions.

The Food and Drug Administration and the United States Department of Agriculture are the agencies responsible for the regulation of nutrition labeling. These agencies ensure that any food to which a nutrient is added or that claims to be nutritional must have a nutrition label. Any additional information is voluntary.

Nutrition label information should be based on the U.S. recommended dietary allowances (USRDAs) that were developed and simplified from the recommended dietary allowances (RDAs). The USRDAs are based on sex and age. The amounts stated on a label should be the recommended amounts for normal healthy adults and children over four years of age, as stated by the USRDA. When reading nutrition labels and comparing different products, look for the following:

- Nutritional information: nutrition information should be expressed in percentage of the USRDA for protein, thiamin, riboflavin, niacin, vitamin A, vitamin C, calcium, and iron. In addition, it should include the number of calories and the grams of protein, fat, and carbohydrates per serving.
- Dietary supplement claims: when intended as dietary supplements, products should be fortified with 50 percent or more of the USRDAs for various vitamins and minerals.
- Enriched claims: enriched means that any vitamins and minerals lost during processing are added to the product.

Another government agency, the **Federal Trade Commission (FTC),** is responsible for dealing with false claims made by any company. If consumers believe that a nutrition claim is false, they should contact the FTC. Due to FTC actions, many companies have been forced to change their advertising or labeling to represent a product more accurately.

The **Fair Packaging and Labeling Act of 1967** ensures that consumers are provided with accurate label information on a food's quality and content. The act requires a food product's manufacturer, distributor, or packer to include the following items on the label:

- The name and address of the manufacturer, distributor, or packer
- The name of the food

Examples are enriched bread and cereal products.

- Fortification claims: fortified means that any vitamins and minerals lost during processing have been replaced in the product and that additional vitamins and minerals are incorporated into the product.
- Low-calorie claims: low-calorie products should contain no more than 40 calories per serving.
- Reduced-calorie claims: reduced-calorie products should contain one-third the calories of comparable products.
- Sugar free claims: sugar free products should not contain sucrose (table sugar) but may contain natural or artificial sweeteners.
- Low sodium claims: sodium content should be stated as *very low sodium*—less than 35 milligrams per serving—*low sodium*—less than 140 milligrams per serving—or *reduced sodium*—reduced from the normal amount by 75 percent.
- Date: although the length of storage is important to the nutritional quality of food products, dating packages is a voluntary service. There are several voluntary methods for dating products, and each one has a different meaning.
    - Pull or sell date: last date a product can be sold
    - Expiration date: last date a product can be eaten
    - Freshness date: date until which freshness is ensured by the manufacturer or packer
    - Pack date: date the product was processed or packaged; tells the age of the product

All nutritional labels must follow standard regulations and include the name and address of the manufacturer or packer, the amount (count, size, and weight), and a list of ingredients, with the ingredient making up the highest percentage of the food stated first, and so on in order.

- The net content in terms of weight, measure, or count
- The name of all ingredients, listed in order, with the ingredient making up the highest percentage of the food stated first

The United States Department of Agriculture (USDA) grades and inspects the quality and condition of agricultural commodities and foods. Meat, poultry, and other processed foods are inspected for wholesomeness. This means that items stamped as *USDA inspected* have been slaughtered or processed under sanitary conditions. Congress has passed several acts establishing the USDA's power to determine the quality of food products. Quality grading is discussed in Chapters 15 through 18, which discuss different food products.

ILLUSTRATION 13–2 The United States Department of Agriculture (USDA) grades and inspects the quality and condition of agricultural commodities and foods.

In addition to a knowledge of federal and state regulations, purchasing agents must have the ability to develop purchasing procedures; select vendors; negotiate food, supply, and service contracts; and monitor deliveries by using a well-organized inventory system. Figure 13–1 illustrates the steps necessary for purchasing, receiving, and storing food and supplies.

Purchasing Procedures

There are several purchasing procedures from which a manager can choose. These procedures include purchasing by bid, fixed bid, weekly bid, or contract.

Bids and Contracts

Bid purchasing is done through price quotations received from different vendors. Price quotations are based on written or verbal product specifications requested by a buyer. The lowest bid is usually accepted by the buyer. A bid request is used by a buyer to solicit bids or prices on a specific item from a prospective vendor or supplier. The date for closing a bid, which guarantees the quoted prices, should be stated on the bid form. Bids are submitted to the buyer in an unmarked, sealed envelope and should only be opened at a designated time.

The **fixed bid** method of purchasing consists of providing written specifications for a bid on an item that is needed in large quantities for a long period of time. The price remains the same for the period agreed upon by the buyer and the vendor whose bid is accepted.

Weekly bid, or weekly quotation, buying is done by telephone and then followed by a written confirmation after a bid is accepted. This method is commonly used to purchase items such as bread, milk, eggs, and fresh produce that need to be ordered in small quantities.

A **contract** is a written agreement between a buyer and a vendor or supplier. It must bear the signatures of both parties in order to be legally enforceable. A well-written contract should include the offer made by the purchaser, the vendor's acceptance, and the price, time, and date of delivery. The vendor is then responsible for delivering the goods by the date indicated in the contract. Failure to meet the delivery date makes the vendor liable for damages if the buyer chooses to take legal action. Once the

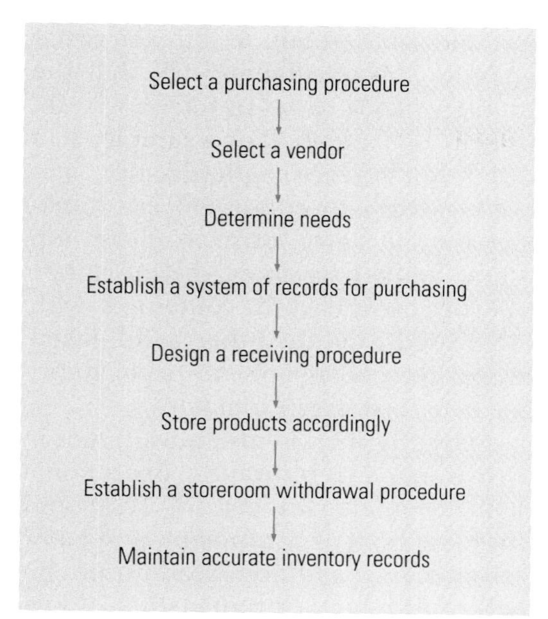

FIGURE 13–1 Steps in the Purchasing, Receiving, and Storing Process

goods have been received (or upon receipt), the buyer must pay the agreed-upon amount. Contracts should be awarded to the most responsible bidder with the lowest price for the quality and service.

Specifications are included with all types of bid or contract requests. A **specification** is a statement describing the quality, characteristics, and quantity of a requested product, including allowable variations. A specification should be clearly written and should include the name brand, federal grade, size, weight, and unit prices on which the final price is established.

Methods of Purchasing

For independent purchasing, each department within a large food service institution bids and purchases its own supplies, or the owner of a small business serves as his or her own buyer. Central purchasing is handled by a special purchasing staff representing all the departments of a large institution. Group purchasing occurs when several departments or institutions agree on certain product specifications and group together for large-volume bid purchasing. This method allows the group to purchase products at the lowest possible price. The purchased food or supplies are divided among the group members according to preset quantities.

Selecting Vendors

Finding vendors that will supply the items needed to operate efficiently is an important task for food service managers. The key factors behind successful relations between vendors and food service managers are product quality, available quantity, and dependable service. A survey of vendors in the surrounding area is a good way for a food service operation to begin searching for vendors that fit its needs. Records of past performance, if available, are the most reliable source to consult when choosing vendors. When a manager is just starting out, however, it is helpful to have opinions and recommendations from personal contacts. Salespeople are another good source of information for vendor selection. If wisely questioned, salespeople, although they are representing the industry for which they work, can offer inside information that may be helpful. Telephone directories can also be used to locate vendors. After all possible vendors have been explored, management must narrow down the group by comparing vendors on product quality and quantity, dependability, and price.

Handling Salespeople

Visits with salespeople must be productive, professional, free of distractions, and should promote good public relations. Salespeople must never be invited to do business during an operation's busy hours, when managers are heavily involved with various responsibilities. They should instead come during the "slow hours," such as afternoons or midmornings, and must arrive on time. Managers should have their order ready so that the visit will be as short as possible, and they should avoid purchasing items suggested by a salesperson for which they have no use. Personal gifts and favors offered by salespeople must never be accepted. Later, a salesperson may expect a favor in return.

Determining Needs

A restaurant's menu is the best guide a manager can use to determine what needs to be purchased. The recipe file is the next item to consult. A study of the recipes used to prepare menu items will help a manager or buyer determine what foods to buy, in what quantity and quality, and whether the product needs to be purchased fresh, frozen, or canned. For example, if a restaurant offers spaghetti sauce, it is not necessary to purchase fresh tomatoes for the sauce, or even to use top grade canned tomatoes, because they will be cooked and combined with other foods. On the other hand, when preparing a salad, a buyer needs to order firm, red, fresh tomatoes.

Selecting food items and supplies is based on a restaurant's operational style, clientele, budget for food and supplies, type of equipment, and labor resources. In addition, reading the labels of purchased products helps buyers select more knowledgeably in the future. When purchasing canned products, for example, buyers should note special labels, such as *dietetic, diabetic, sugar added, enriched,* and *number*

ILLUSTRATION 13–3 When preparing a salad, a buyer needs to order firm, red, fresh tomatoes.

of pieces, as well as foods that contain monosodium glutamate.

Whenever possible, managers or buyers should purchase foods in large amounts, especially seasonal or sale items. They should purchase in large quantities, however, only if the entire amount can be used in a reasonable period of time without compromising quality. Buyers must consider the amount of available storage space. This includes storage for frozen, refrigerated, and dry foods, in addition to cleaning supplies. Inventory should never be overstocked. An overstocked inventory ties up capital funds, may cause deterioration of food products, and encourages losses from theft. On the other hand, if products are purchased in amounts too close to those needed, an operation may run short and be forced to make an emergency order, which may or may not arrive in time, depending on transportation availability and how much a vendor has in stock.

Purchasing Records

The three types of record-keeping forms commonly used during the purchasing process are the purchase requisition, the purchase order, and the invoice. A **purchase requisition** is used to request food and supply items. A **purchase order** states specific information regarding the items to be purchased, and it also serves as an agreement between a buyer and a vendor. At least three copies should be available: one for accounting purposes, one for the receiving depart-

MATH SCREEN

When preparing purchase orders, food service managers determine the total cost of an item by multiplying the quantity purchased by the price per unit. Use the amounts listed in the following table to calculate answers to the problems in this Math Screen.

Item	Quantity	Unit	Unit Price
Oranges	15	case	$9.15
Tomatoes	7	lug	9.75
Lettuce	3	case	4.75
Cucumbers	2	dozen	1.95

Study this example to see how to calculate an item's total cost: What is the total cost of the oranges?

Quantity = 15

Unit price = $9.15

15 × $9.15 = $137.25

Test your math skills with the following problems:

1. What is the total cost for the tomatoes?
2. What is the total cost for the lettuce?
3. What is the total cost for the cucumbers?
4. What is the total cost of all the produce items in the table?

PURCHASE ORDER

NAME OF INSTITUTION ______________________

DEPARTMENT ______________________

ADDRESS ______________________

____

TO ______________________

____

____

DATE ______________________

PURCHASE ORDER # ______________________

(Refer to above number on all invoice forms)

REQUISITION # ______________________

DATE NEEDED ______________________

Item	Description	Quantity	Unit	Unit Price	Total Cost

TRANSPORTATION ______________________ TERMS ______________________

APPROVED BY ______________________

TITLE ______________________

SPECIAL INSTRUCTIONS: Prepare in triplicate for the vendor, accounting office, and manager.

FIGURE 13–2 Purchase Order Form

ment, and one for the department that requisitioned the order. Figure 13–2 illustrates an example of a purchase order. The form may vary in format from one organization to another.

An **invoice** is a form written by the vendor. It contains information such as item description, quantity, and price. When requested items are received by a buyer, the invoice is compared to the purchase order for accuracy. Any discrepancies should be checked and immediately noted on the invoice by the receiving personnel and initialed by the delivery person representing the vendor. The invoice and purchase order are the two forms used to secure payment.

RECEIVING AND STORING

In food service establishments, approximately 40 percent of annual revenues is spent on food and supplies. To protect this investment from theft and waste, managers must assign responsible employees to receive and control inventory. Economic gains from good purchasing procedures may be wasted if an organization's receiving and storing practices are poor.

Receiving Procedures

Efficient receiving requires adequate space in the receiving area as well as necessary equipment, qualified receiving personnel, and appropriate storage space for food and supplies.

Incoming orders are unloaded in the receiving area, which also serves as an entrance for salespeople and food service personnel. The manager's office should be located near the receiving area, and it should have a clear view of the entrance so that the manager can monitor deliveries, salespeople, and personnel. Time and money are saved if the receiving area is designed to promote prompt storage. Storage space located near the receiving area also prevents delivery personnel and salespeople from passing through the production area of the kitchen.

Receiving Personnel

One or several employees may receive incoming deliveries, depending on the size of a food service facility. Their responsibilities include receiving and storing delivered items, withdrawing food and supplies from storerooms, and controlling inventory levels. If possible, it is wise to assign the responsibility of scheduling and receiving deliveries to the same employee who purchases items. This assignment ensures that ordered items are accepted or rejected with confidence and that purchasing errors are minimized. In addition, it can prevent conflicts that often arise between employees when responsibilities are split or unclear. Deliveries should be scheduled during a time when the employee or employees responsible for receiving them are less occupied with other duties and have the time to check each item carefully for quality and accuracy. Receiving personnel should always be knowledgeable in product quality and specifications, and in how to check and evaluate an item.

Storing Needs

After an item is received, it should be marked with the date of receipt, the weight or count when received, and the vendor's name. It should then be placed in proper storage as soon as

ILLUSTRATION 13–4 Deliveries should be scheduled during a time when the employee or employees responsible for receiving them are less occupied with other duties and have the time to check each item carefully for quality and accuracy.

possible. Perishable food items that require freezing or refrigeration should be stored immediately. Prompt storage prevents or slows deterioration and spoilage, maintains quality, and ensures food safety.

Food and supplies are usually stored at their point of first use, with previously purchased items in the front of shelves and newly received items in the back. This placement prevents items from being forgotten in the back of shelves for long periods, causing spoilage. The rule to remember when storing food and supplies is "first in, first out," or FIFO.

Storage areas should be located close to appropriate working areas to minimize the labor involved in transporting supplies. In large food service kitchens, mobile storage units and carts make storage easy because they can be brought to the receiving area, loaded, and wheeled directly into a walk-in freezer, refrigerator, or dry storage area.

Perishable Food Storage

Perishable foods must be stored promptly in refrigerators or freezers to preserve their quality and nutritional value and to prevent the growth of bacteria that cause foodborne illness. In large food service operations, separate refrigeration or freezer units may be available for meat and poultry, fish, dairy products, and vegetables and fruits. These separate units are ideal because they minimize the absorption of different odors given off by various foods. In smaller restaurants, however, separate units are not feasible because it is costly to purchase, maintain, and operate equipment unless it is used to its full capacity.

Dry Storage

The dry storage area should be designed with enough space for all dry, nonperishable foods, paper products, and miscellaneous items. Cleaning supplies should always be stored separately to prevent the accidental use of these products as food. Dry air, a temperature below 65°F, adequate ventilation, appropriate light, and efficient organization are five factors that contribute to a good dry storage area.

Inventory Control

Regular inventories can help a buyer establish the amount of food or supplies he or she needs to order, determine what items are available for use, recognize pilferage, assist in future forecasting, and identify price changes that may affect purchasing decisions. Food service managers use several methods to track and control the valuable goods in storage areas. Two of the most commonly used methods are perpetual and physical inventories.

Perpetual Inventories

A **perpetual inventory** record is a card or sheet on which employees note the receipt, withdrawal, amount, and cost of items in cold and dry storage areas.

At any given time, a manager or buyer can glance at the perpetual inventory records to determine the quantity and value of all inventoried items. A perpetual inventory record form should contain space for an item's description and specifications, size, unit, quantity, and unit cost, as well as adjacent columns for noting when and in what amounts items are received and withdrawn. An example of a perpetual inventory record form is illustrated in Figure 13-3.

Physical Inventories

A **physical inventory** record is a monthly or quarterly report on all the items physically in a facility's storage areas. The employee or employees responsible for this record should inventory items systematically and, if needed, rearrange items as changes occur. A physical inventory record form may be designed to reflect the order in which items are stored. The form should include the date and dollar amounts recorded during the last inventory, item descriptions and sizes, actual counts of cans and cases on hand, unit costs, the total cost of all items, the date, and the signature of the employee taking the inventory.

Controlling Pilferage

Pilferage refers to the repeated stealing of small amounts of food or other items. Pilferage problems can be con-

FIGURE 13–3 Perpetual Inventory Record Form

PERPETUAL INVENTORY RECORD

ITEM: UNIT/SIZE:
DESCRIPTION & SPECIFICATIONS: QUANTITY/SIZE:
VENDOR: COST/UNIT:

Date	Amount Received	Amount Withdrawn	Employee Withdrawing	Unit on Hand	Cost per Unit	Total Cost

trolled by keeping all storage areas locked when food production is finished, by permitting only one employee to have a key to the storage areas, and by requiring employees to record when they add to or withdraw from stored items. Expensive foods, such as caviar, canned anchovies, salmon, coffee, and wines, should be kept in a separate storage area that is locked at all times.

CAREER PROFILE

POSITION: SCHOOL FOOD SERVICE WAREHOUSE COORDINATOR

Qualifications:

- Good verbal and written communication skills
- Good math computation skills
- Three-years experience in warehouse shipping and receiving or a combination of training, education, and experience to meet the minimum qualifications

Responsibilities:

- Coordinates all the operations of a food service central warehouse
- Monitors the activities of delivery workers and storekeepers
- Arranges for the storage of food items and supplies
- Ensures that all food items and supplies are stored according to local health department regulations
- Establishes and assigns delivery routes and evaluates their effectiveness, making changes as necessary
- Investigates complaints from the schools regarding food items not received and arranges for the delivery of needed items
- Inspects incoming shipments and acknowledges receipt of items as specified on the receiving invoice and purchase order
- Arranges for the repair of all equipment used to transport food items
- Conducts monthly inventories of all government commodities
- Conducts annual physical inventories of all warehouse items

SUMMARY

Food service managers and employees can successfully purchase, receive, and store foods and supplies if they are familiar with the current food market, are able to forecast future needs, and are well trained in purchasing and receiving procedures. In particular, they should know how to use various record-keeping tools, such as bid forms, purchase orders, requisitions, and different types of inventory forms. These tools help managers and employees to maintain the quality and quantity of products and to control costs.

Purchasing agents must know and understand all federal and state regulations having to do with food processing and transportation. In addition, they must be able to use different types of bids and contracts, select and deal with vendors and salespeople, and determine their organizations' needs.

Well-organized receiving procedures save time and money. The receiving area should be designed to enable prompt storage. Receiving personnel should be trained to efficiently compare a purchase order with the invoice accompanying delivered items.

All received items should be marked with their date of receipt. Perishable items need to be stored immediately. Food service personnel should also keep in mind the "first in, first out" guideline—newly received items should be stored behind previously purchased items. Inventory records, whether done by the perpetual or the physical inventory method, need to be regularly and carefully maintained.

QUESTIONS

1. Why is it important for a buyer to know the state and federal regulations concerning the production, processing, and transportation of food products?
2. What government agency is responsible for enforcing the Food, Drug, and Cosmetic Act of 1938?
3. What information does the Fair Packaging and Labeling Act of 1967 require on food labels?
4. What method of purchasing involves a buyer providing a group of vendors with written specifications for a price quotation on a particular item?
5. What type of purchasing is usually done by telephone, followed by a written confirmation after acceptance of the bid?
6. What written or verbal statement, given by a buyer to a vendor, describes the quality, characteristics, and quantity of a requested product?
7. What are the differences between independent purchasing, central purchasing, and group purchasing?
8. What four sources should buyers consult to locate vendors?
9. What necessary information does a buyer need before placing an order with a vendor?
10. What form provides specific information regarding the items to be purchased, and also serves as an agreement between a buyer and a vendor?
11. What are the differences between the two most commonly used methods of inventory control?
12. What does the term *first in, first out* mean?
13. What five factors contribute to good dry storage areas?

ACTIVITIES

Activity One Inspect the labels of canned products in your home or classroom for the name and address of the manufacturer, as well as the item's weight, measure, count, and nutritional information. List the ingredients of these products, with the ingredient making up the highest percentage of the product stated first.

Activity Two Ask your instructor for a standardized recipe. Practice ordering the ingredients in this recipe by filling out a copy of the purchase order shown in Figure 13–2 on page 170.

PART SIX

QUANTITY FOOD PRODUCTION

14
FOOD PREPARATION

VOCABULARY

dry heat cooking
larding
barding
deglassing
pounding
sautéing
moist heat cooking
blending
creaming
cutting in
folding
kneading
rounding
sifting
whipping

OBJECTIVES

After studying this chapter you should be able to do the following:

- Describe the two methods of cooking and the different techniques used in each one
- Demonstrate the different cutting techniques, using the knife appropriate to each
- Identify the different mixing techniques and their purposes
- Convert temperatures from Fahrenheit to centigrade and from centigrade to Fahrenheit

Selecting which method or technique to use in food preparation depends on the size, quantity, and quality of the food being cooked, as well as on the availability of time and the desired outcome. Inappropriate cooking techniques can ruin food. Overcooked meat, for example, becomes dry and loses its flavor. If meat is properly cooked, it will be juicy, flavorful, tender, and nutritious.

People eat with their eyes first. Therefore, food must be prepared and served to look appetizing. Shapes and textures play an important role in food preparation. Knives, which can help shape foods, are the most important hand tools in a kitchen. A chef or prep cook is usually judged by her or his ability to handle knives and various cutting techniques.

There are several mixing techniques used in food preparation that determine the appearance and quality of a cooked product. In addition, all cooks should be aware of the nutritional content of foods and the effects that different cooking methods may have on a food's nutritional value.

COOKING METHODS

Many different techniques are used in cooking. There are, however, only two cooking methods: the dry-cooking method and the moist-cooking method. Within each method different techniques are used.

Dry Heat Cooking

In the **dry heat cooking** method, heat is transferred by radiant energy and air. When radiant energy is used, heat flows from a substance of higher temperature to a substance of lower temperature. Energy sources for dry heat cooking include red electric heating elements, gas flames, hot coals, and flames from wood and charcoal. In charcoal cooking, for example, the high-temperature charcoals produce radiant energy, which is absorbed by the lower-temperature food. Dry heat cooking does not require liquid. Roasting, broiling, pan broiling, grill-

MATH SCREEN

To convert degrees Fahrenheit to degrees centigrade, follow this formula:.

$°C = 5/9 \times (°F - 32)$

Example: What is the centigrade temperature for 250°F?

$°C = 5/9 \times (250 - 32)$

$°C = 5/9 \times (218)$

$°C = 1090/9$

$°C = 121$

To convert degrees centigrade to degrees Fahrenheit, follow this formula:

$°F = (C° \times 9/5) + 32$

Example: What is the Fahrenheit temperature for 100°C?

$°F = (100 \times 9/5) + 32$

$°F = 900/5 + 32$

$°F = 180 + 32$

$°F = 212$

Test your math skills:

1. What is the Fahrenheit temperature for 40°C?
2. What is the centigrade temperature for 95°F?

ing, pan frying, stir frying, and deep-fat frying are various methods of dry heat cooking.

Roasting

Roasting is done in an oven. To cook a meat roast, the roast is placed in a pan large enough to hold the meat and its juices. A meat roast thermometer is usually placed in the center of a roast to ensure desired doneness. Table 14–1 shows approximate roasting times for various kinds and cuts of meat.

Roasting works best when tender cuts of meat are used. Prime rib, whole-cut sirloin, leg of lamb, chicken, and turkey are popular menu items that are roasted in many American restaurants. Less tender cuts of meat used in roasting require some pre-preparation. For example, if a meat is extra lean, larding or barding the meat may be necessary to make it more tender. In **larding,** strips of fat are pushed into the inner part of the meat using a larding needle, while in **barding,** fat is placed on top of the meat and allowed to soak through it during cooking. Both of these methods produce a more tender roast.

Nutrition-conscious cooks, however, will usually remove excess fat from chicken and meat. Beef and pork cuts are especially high in fat; trimming these meats carefully helps to cut down on calories and cholesterol. Cooks can keep meats tender by brushing them lightly with vegetable oil, which is low in cholesterol (but high in calories).

Broiling

Broiling is cooking by directly exposing food to a heat source. The heat may come from charcoal, an electric

TABLE 14–1 Roasting Time Chart (Oven Temperatures from 300°F to 325°F)

TYPE OF ROAST	WEIGHT (IN POUNDS)	DESIRED DONENESS	MEAT THERMOMETER READING	APPROXIMATE COOKING TIME (IN MINUTES PER POUND)
Prime rib	4–7	Rare	140°F	26–32
		Medium	160°F	34–38
		Well done	170°F	40–42
Tenderloin roast	4–6	Rare	140°F	45–60
Boneless round roast	4–6	Rare	140°F	25
		Medium	160°F	35
		Well done	170°F	45
Meat loaf	2–3	Medium	160°F	30
Chicken	2–3	Well done	170°F	18
Turkey	8–12	Well done	170°F	25
Fish (whole salmon)	4–6	Medium	160°F	12
Lamb (leg)	5–6	Well done	170°F	45
Pork (boneless)	4–6	Well done	170°F	45

burner, or a gas flame. The cooking time, temperature, and distance of the food from the heat source are critical elements of broiling. Doneness and browning can be controlled by moving the grids supporting the food either closer to or further away from the heat source.

Broiling time depends on the type, grade, thickness, and desired doneness of a food. Vegetables commonly used in broiling are onions, green peppers, cherry tomatoes, zucchini, and mushrooms. Popular broiled entrées include steaks, chops, chicken, and ribs. While broiling, foods can be brushed with melted fat, barbecue sauce, marinades, and other seasonings for a desired flavor.

Thicker pieces of meat need to be broiled further away from the heat source and allowed to cook longer. Thinner pieces can be placed closer to the heat source and cooked for a shorter time period.

Pork and chicken products should be broiled five inches away from the heat source and allowed to cook until well done. Cooks should turn a food item, using tongs, only once for each side during broiling. Frequent turning or piercing with a fork drains juices and dries out the food.

A probe-type meat thermometer may be used to check the doneness of thick products. Rare steaks are broiled to internal temperatures of 140°F, medium steaks to 160°F, and well-done steaks to 170°F.

ILLUSTRATION 14–1 A probe-type meat thermometer may be used to check the doneness of thick products.

Pan Broiling

Pan broiling is used to broil thin and, often, lean cuts of meat. The meat is placed in an uncovered frying pan, without added fat, and cooked on the surface units of a range. It is browned on both sides in its own fat, and the excess fat is usually poured off. After removing the excess fat, the brown particles and juices that remain in the pan may be deglassed to make gravies or sauces. **Deglassing** means adding water to a pan in which meat has been roasted, sautéed, or panbroiled to dissolve the juices that have dried on the bottom and sides of the pan.

Grilling

The grilling process consists of placing a cut of meat directly on a hot pan or on the solid surface of a griddle or grill. Typically, no additional fat is used for cooking. Hamburgers and steaks are usually grilled, especially in fast-food restaurants.

If hamburgers are being grilled, they should be about 18 to 25 percent fat. Most meats contain enough fat to be grilled.

A griddle can be cleaned with a pumice stone. A pumice stone is a spongy, light rock, which, when rubbed over a griddle, will clean it without scratching.

Pan Frying

In pan frying, a food is cooked in small amounts of hot fat. The food being fried may be coated with bread crumbs or flour. Less expensive, lean cuts of meat are usually pounded with a wooden or metal mallet before pan frying. **Pounding** a meat shortens its fibers and produces a more tender product. The sharp points on the surface of a mallet break down a meat's connective tissue fibers.

Pan frying can be used to cook a variety of foods. Meats, fish, poultry, vegetables, fruits, and batter foods, such as fruit fritters and eggs, are often panfried.

Stir Frying

Stir frying originated in the Orient and is done with an uncovered wok and hot oil. A wok is a heavy, metal pan with sloped sides and a rounded bottom. This method of cooking has been popular during the last decade and is used to cook a variety of meat and vegetable dishes. Food is stir-fried at high temperatures, ranging from 375°F to 400°F, for a short period of time; meats used for stir frying, therefore, must be tender. When stir-frying, meats and vegetables should be cut into small pieces so that they cook rapidly and uniformly. The finished product must be tender and crisp. **Sautéing** is the same as stir frying except that less oil is used. Sometimes

ILLUSTRATION 14–2 Stir frying originated in the Orient and is done with an uncovered wok and hot oil. This method of cooking has been popular during the last decade and is used to cook a variety of meat and vegetable dishes.

the terms *pan frying* and *sautéing* are used to mean the same thing.

Deep-Fat Frying

Deep-fat frying, also called french frying, involves completely submerging food in hot vegetable oil. The thermostat on a deep-fat fryer is usually set between 325°F and 350°F. This method of cooking is very popular because it cooks food to a crisp state that most people find very tasteful.

Breaded food items are excellent when deep-fried; however, they are also high in calories. Foods that can be deep-fried include meats, fish, chicken, onions, zucchini, mushrooms, eggplant, and, of course, the popular potato.

For best results, a deep-fat fryer's oil should be filtered and changed frequently. In addition, the oil must have a high smoke point. The smoke point is the temperature at which a fat disintegrates and begins to smoke.

Moist Heat Cooking

Moist heat cooking requires moist air and water. To cook with moist heat, cooks can use a closed pan, with or without added liquid; can use a pressure cooker; can add liquid to an uncovered pan; or can steam food on a rack above a liquid.

Meats cooked with moist heat usually come from animal parts that are exercised and, thus, tough. These meats usually require more effort to cook but are more flavorful than tender cuts. Moist-cooking methods include boiling, braising, stewing, steaming, and pressure cooking.

Boiling

When boiling, food is placed in a saucepan or a stock pot, completely covered with water, and cooked on a range top. This method cooks food evenly without turning. Herbs and seasonings may be added to flavor the food. The pan may be covered or uncovered, depending on the item being prepared.

Boiling is used to cook corned beef, ham, pastrami, chicken, turkey, eggs, and a variety of vegetables, soups, and stocks. These foods are usually simmered at 200°F after the water reaches the boiling point of 212°F. When stocks and soups are being prepared, the pan should be left uncovered, and residue floating on top of the liquid should be removed with a ladle. Skimming this residue eliminates possible cloudiness and produces a clear and thick product. Leaving the pan uncovered permits the water to evaporate and the soup or stock to become concentrated.

Vegetables boiled in large amounts of water lose some of their important vitamins. These water-soluble vitamins are the B vitamins and vitamin C. To preserve their nutrient content, vegetables should be boiled in small amounts of water until tender. The water should be brought to a boil before adding the vegetables. Strong-flavored, green, and white vegetables

NUTRITION NOTE AVOIDING NUTRIENT LOSS

Care must be taken during food preparation and cooking to minimize the loss of valuable nutrients. Vegetables and fruits are rich in vitamins B and C as well as many minerals that are necessary for good health. Soaking these foods in water during cooking or preparation reduces their vitamin and mineral content.

Some loss of nutrients is unavoidable. It can, however, be reduced substantially by following these tips:

1. Whenever possible, avoid soaking food.
2. Use minimal amounts of water to cook food.
3. Shorten cooking time by covering pans to retain heat.
4. Cook foods to a tender-crisp stage.
5. Use leftover liquids in soups and sauces.

Heat destroys the B vitamins and vitamin C. Thiamin, one of the B vitamins, is especially subject to destruction by dry heat, such as the heat used to toast bread. Heat destruction of vitamins can be reduced by cooking vegetables to a tender-crisp stage and by avoiding unnecessary toasting.

The B and C vitamins are air soluble. When it is necessary to peel or cut fruits and vegetables during food preparation, delay this process as long as possible to prevent overexposure to the air.

---

should be boiled uncovered to preserve their color and flavor and to permit volatile acids to escape.

Braising

Braising is cooking food in its own juices, after it has been browned, with a small amount of added liquid. Foods can be braised in an oven or on a range top, covered or uncovered. Covering a braising pot allows food to cook more evenly.

Braising is also called fricasseeing. To fricassee is to cook meat in its own gravy. When braising meats, they are browned first to develop better color and flavor. Braising is done at a low temperature, between 250°F and 275°F, which results in a higher yield. This technique is used to cook less tender cuts of meat, vegetables, and many other foods. Braising is desirable because it preserves the nutrient quality of food.

Stewing

Stewing is similar to braising, except that the meat and vegetables are cut into small, uniform pieces, and more liquid is used. In stewing, the stock or the water should cover the food. The pot is covered throughout cooking, and the food should be cooked at a low temperature of 250°F, which reduces shrinkage and increases the tenderness of the meat.

When vegetables are included with stew meat, they should be added during the last 20 minutes of cooking to prevent overcooking. Different vegetables require more or less cooking, depending on the type. Each vegetable should be cooked separately and then added to the stew near the end of the cooking process.

Stewing is usually done on a range top. It can, however, be done in an oven.

Steaming

Steaming is cooking food without placing it in direct contact with the water that provides the steam. To steam food, it is placed in a perforated basket, and the basket is fitted into a covered saucepan with a small amount of water under the basket. The water simmers and produces steam, which cooks the food and drops back down into the water. Steaming is used to cook meats, fish, fruits, and, most commonly, vegetables.

Pressure Cooking

Pressure cooking is the combination of boiling and steaming with superheated water under pressure. Vegetables are usually cooked with 5 pounds of pressure and meat with 15 pounds of pressure. The exact cooking time depends on the cut, size, and type of food being prepared. Although pressure cooking is appropriate for a variety of different foods, it is most commonly used for vegetable cookery, as shown in Table 14–2. Pressure cooking helps to preserve vegetables' nutritional value.

TABLE 14–2 Pressure-Cooking Fresh Vegetables (with 5 Pounds of Pressure)

VEGETABLE	AMOUNT OF VEGETABLE	APPROXIMATE COOKING TIME
Green beans	10 lbs	25 minutes
Beets	10 lbs	1 hour
Broccoli	10 lbs	10 minutes
Cabbage	10 lbs	20 minutes
Carrots	10 lbs	25 minutes
Cauliflower	10 lbs	15 minutes
Celery	10 lbs	15 minutes
Corn on the cob	10 lbs	10 minutes
Kale	10 lbs	30 minutes
Onions	10 lbs	25 minutes
Potatoes	10 lbs	40 minutes
Spinach	10 lbs	10 minutes
Squash	10 lbs	20 minutes
Sweet potatoes	10 lbs	30 minutes
Turnips	10 lbs	15 minutes

CUTTING TECHNIQUES

Although commercial food service establishments use many different cutting tools in food preparation, the majority of cutting is done by hand. Hand tools are especially needed for cutting delicate fruits and vegetables. Cutting machines are too harsh for many foods; they damage the food's texture and cannot produce many desired shapes. In addition, when small amounts of an item are needed, it is more efficient to use knives or other hand tools. Time is saved because cutting machines need more labor hours to clean. Various cutting techniques are described in Table 14–3.

ILLUSTRATION 14–4 Mincing cuts food into the smallest possible size with a French knife. Parsley is often minced.

ILLUSTRATION 14–3 Dicing is used to cut food into tiny cubes approximately one-fourth square inch in size. Performed with a French knife, this technique is commonly used on onions.

ILLUSTRATION 14–5 Slicing cuts food into layers of uniform thickness. A French knife is used to slice fruits and vegetables, such as tomatoes.

TABLE 14–3 Cutting Techniques

Cutting This technique is used to cut food into many uniformly small pieces using a French knife.
Dicing Dicing is used to cut food into tiny cubes approximately one-fourth square inch in size. Performed with a French knife, this technique is commonly used on onions and celery.
Mincing Mincing cuts food into the smallest possible size with a French knife. Parsley, herbs, and vegetables are often minced.
Slicing Slicing cuts food into layers of uniform thickness. A French knife or a slicing knife is used to slice meats and fruit.
Shredding Shredding is used to cut food into thin strips or bits that are uniform in size and shape. A French knife is used to shred lettuce, cabbage, carrots, and cold meats.
Grating Grating cuts food into coarse pieces. To grate food, it is rubbed against the sharp, perforated edges of a grater. A boxed grater has four different cutting sides and is commonly used to grate cheese and onions.

MIXING TECHNIQUES

Mixing is the process of combining two or more ingredients into one uniform substance. Mixing can be done with hand tools, mixing and blending machines, or other food processors. Each mixing technique is unique for the desired outcome of the food being prepared. For example, blending is the correct technique to use when preparing muffins. If the muffin batter is overmixed, the muffins will bake with an uneven surface and hollow interior. Table 14–4 defines the mixing techniques used in food preparation.

TABLE 14–4 Mixing Techniques

Blending **Blending** is combining two or more ingredients thoroughly with a spoon. All the ingredients may be combined and blended at one time. This technique is commonly used to make muffins.
Creaming **Creaming** incorporates air into a mixture to make it smooth and creamy. It is often used when combining sugar and a shortening.
Cutting in **Cutting in** blends two ingredients together by rubbing them with the palms of the hands or by using a pastry cutter.
Folding **Folding** mixes ingredients by gently rotating a spoon or spatula from the top of a mixture to the bottom and from the bottom back up.
Kneading **Kneading** is folding, stretching, and pressing dough to push the air out of it. To knead properly, press the dough with the palms of the hands while stretching and folding it at the same time.
Rounding **Rounding** is shaping dough on a floured surface to seal its ends. The dough should be rolled with the palms of the hands in a forward motion while the edges are sealed.
Sifting **Sifting** means passing dry ingredients through a sieve or flour sifter to make a lighter product.
Whipping **Whipping** mixes ingredients by beating them at high speed with a mixing machine. This process incorporates air into a substance and makes it light and firm. The substance should be whipped to a stiff consistency.

CAREER PROFILE
POSITION: RESTAURANT COOK

Qualifications:

- High school education or equivalent
- Education in food service or cooking desired
- One year of experience in commercial cooking
- Good communication skills

Responsibilities:

- Works as a team member with other cooks to prepare a restaurant's menu items
- Prepares, seasons, and cooks large quantities of soups, meats, vegetables, desserts, and other foods
- Maintains standards of quality, safety, and sanitation
- Estimates the amount of food needed and orders it from suppliers
- Adjusts thermostat controls to regulate the temperatures of ovens, broilers, grills, roasters, and steam kettles
- Measures and mixes ingredients according to recipe specifications
- Uses a variety of kitchen tools and equipment, such as blenders, mixers, grinders, and slicers, during food preparation
- Carves meats and portions food on serving plates
- May wash, peel, cut, and shred vegetables and fruits
- May butcher chicken, fish, and shellfish
- May price menu items
- Supervises the cook's helper and the prep cook
- Reports unsafe working conditions

SUMMARY

The different methods and techniques used in cooking make foods more tender, juicy, flavorful, appetizing, and nutritious. A chosen preparation technique should enhance these characteristics in all foods.

Dry heat and moist heat cooking are the two basic cooking methods. Dry heat cooking techniques include roasting, broiling, pan broiling, grilling, pan frying, stir frying, and deep-fat frying. The major moist heat cooking techniques are boiling, braising, stewing, steaming, and pressure cooking.

Cutting and mixing techniques enhance the appearance of food by creating appealing textures, shapes, and consistencies. Cooks must know how to use hand tools, such as knives and graters, as well as a variety of complex cutting and mixing machines.

Most importantly, food service managers must take appropriate steps to preserve the nutrient content of food, which is often lost during preparation and cooking. Mastering different methods and techniques of food preparation promotes healthful cooking as well as a safe and efficient work environment.

QUESTIONS

1. What is the difference between dry heat cooking and moist heat cooking?
2. What is meant by *larding* and *barding?*
3. How is pan broiling done, and what is its relationship to deglassing?
4. Why is it necessary to pound less tender cuts of meat with a mallet before cooking?
5. What is another name for *stir frying?*
6. How is deep-fat frying done, and what is another name by which this cooking technique can be identified?
7. How is braising done, and by what other name is it known?
8. How is stewing done?
9. How are steam cooking and pressure cooking different? How are these techniques similar?
10. What are four techniques used to cut foods?
11. What are four techniques used to mix food products?

ACTIVITIES

Activity One Ask your teacher to demonstrate the different techniques used to cut vegetables by using the appropriate knife and the techniques listed below. Practice until you can perform all the different cutting techniques.

Vegetable	Cutting Technique
Celery	Chop
Onions	Dice
Tomatoes	Slice
Parsley	Mince
Carrots	Shred

Activity Two Ask your teacher to explain and demonstrate the following mixing techniques:

1. Mixing—use a boxed brownie mix recipe
2. Whipping—beat one pint of whipping cream with one tablespoon of granulated sugar, using an electric mixer on high speed
3. Creaming—Beat two ounces of butter, three ounces of cream cheese, one teaspoon of vanilla, and two cups of sifted confectionary sugar until smooth and creamy

Use tasting spoons to evaluate the three different techniques of mixing (before baking the brownies). Judge each technique by the characteristics that follow, using a rating scale of 1 (poor), 2 (good), or 3 (excellent). Explain reasons for the rating you give each product.

- Color
- Texture (creamy, velvety, smooth)
- Consistency (thick, thin, fluffy)
- Taste

Bake the brownies according to the directions in the box and cool. Meanwhile, refrigerate the whipped cream. The cream cheese frosting can be left at room temperature until the brownies are ready. Top the brownies with whipped cream or cream cheese frosting and evaluate the finished product.

RECIPES

Fried Meatballs

Number of Servings: 24

Prepared Ingredients	Weight	Measure	Method
Ground beef	3 lbs		Combine ground meats, bread (lightly dampened with water or milk), eggs, and the remaining flavoring ingredients. Mix well with hands until all ingredients are blended.
Ground lamb	3 lbs		
Wheat bread, torn in small pieces	¼ lb		
Eggs		4	
Onion, finely chopped		2 cups	Pinch off small amounts from the meat mixture to form 1½-inch round balls approximately 1 ounce in weight. Continue making meatballs until all mixture is used.
Garlic, minced		4 cloves	
Fresh mint, crushed		¼ cup	
Dried oregano, crushed		1 tbsp	Lightly coat meatballs with flour and panfry in hot oil until golden brown.
Dried basil, crushed		1 tsp	
Salt		2 tsp	
Pepper		½ tsp	
Flour, for coating		1 cup	
Olive oil, for frying		1 cup	

Serving Size: 4 meatballs.

Serving Suggestions: Excellent as an appetizer or as a main dish with scalloped potatoes and a tossed green salad on the side.

Baked Halibut

Number of Servings: 24

Prepared Ingredients	Weight	Measure	Method
Halibut steaks	12 lbs	24 8-oz steaks	Preheat oven to 350°F.
Salt		1 tsp	Salt and pepper halibut steaks and coat with olive oil and lemon juice in a large, shallow roasting pan. Add remaining ingredients on top of the fish steaks, making sure that they are evenly spread out.
Pepper		½ tsp	
Lemon juice		1 cup	
Olive oil		½ cup	
Garlic, sliced		6 cloves	Bake for 45 minutes at 350° F.
Tomatoes, crushed		4 16-oz cans	
Onions, sliced	2 lbs		
Fresh parsley, minced		¾ cup	

Serving Size: One 8-ounce steak.

Serving Suggestion: Serve with rice pilaf and a tossed green salad on the side.

15

MEAT, POULTRY, AND FISH

VOCABULARY

marbling
wholesomeness
wholesale cuts
yield grade
retail
Trichinella spiralis
stock
bouillon
consommé
clarification
borscht
vichyssoise
gazpacho
sauce

OBJECTIVES

After studying this chapter you should be able to do the following:

- Identify the grading systems used for quality meat, poultry, and fish products
- Identify the different types of animals bred for human consumption
- Recognize different market forms of meat, poultry, and fish
- Explain the reasons why meat, poultry, and fish are important to a healthful diet
- Describe the ways in which stocks, soups, sauces, and gravies are similar and different

Meat, poultry, and fish are the major components of most daily menu entrées. A complete meal is usually planned around the meat entrée. Meat is the muscle tissue of cattle, sheep, pigs, and other animals. Poultry is the flesh of birds and game such as chicken, turkey, duck, pheasant, and quail. In the United States, turkey is a popular holiday entrée for Thanksgiving and Christmas. Fish includes finfish and shellfish. The popularity of fish dishes has increased tremendously in recent years as people have become more concerned with health. Fish has less fat and cholesterol than red meats or poultry.

Managers at commercial food service establishments should know the characteristics that affect the quality of meat, poultry, and fish, such as the age of an animal. They should also be familiar with the way in which these foods are inspected and graded and with the market forms available. In addition, kitchen staff should know how to use meat, poultry, and fish—a major source of dietary protein—as the main ingredients for many stocks, soups, sauces, and gravies.

MEAT

Meat is the flesh of mammals, such as cattle, hogs, and young sheep, that are bred for human consumption. All animal products are composed of muscle tissue, connective tissue, fat, and bone.

The visible fat that is embedded in the lean tissue of meat is called **marbling.** Factors contributing to the marbling of meat are the age of the animal, the type of feed, and the amount of exercise the animal received. Animals raised with minimal exercise have meat with more marbling, which improves the flavor, tenderness, and juiciness of the meat.

The bone makes up the skeletal structure of an animal and helps identify the age of the animal and the cut of meat. Young animals have soft, pinkish bones, while older animals have bones that are hard and white.

Inspection and Grading

Meat inspection and grading are done by highly trained, qualified USDA inspectors and graders. Food service establishments depend on USDA officials to carefully check and grade the meat they purchase.

Inspection

During the inspection process, meat is carefully looked over and compared to an established set of quality standards. The law requires that all meat products sold must pass inspection for **wholesomeness.** USDA approval for wholesomeness guarantees that purchased meat is from healthy animals slaughtered and processed under sanitary conditions. Meat that has passed federal inspection is stamped with a round, purple mark (see Figure 15–1). This mark indicates the official number assigned to the establishment where the animal was slaughtered and processed. The inspection stamp is placed on the wholesale cuts. The **wholesale cuts** are the large sections of meat into which a carcass is divided.

FIGURE 15–1 Stamp for Wholesomeness

Grading for Quality

USDA graders rate meat according to its quality. The type of grade determines the selling price. Animals of higher quality grades are few and carry higher price tags. USDA Prime is the highest quality grade. This meat contains the greatest degree of marbling. Prime quality meat is generally sold to fine restaurants and hotels. Table 15–1 lists the quality grades given to beef, veal, lamb, and pork. Grading is based on certain characteristics, including the degree of marbling; the meat's color, texture, and finish; and the age of the animal when slaughtered. For pork, the quality grade is actually a yield grade (discussed in the next section) based on the degree of muscling and the degree of fatness. Muscling is judged from very thick to very thin. The grade mark for quality is a shield-shaped symbol with the letters USDA and the grade name (see Figure 15–2). The shield is stamped on wholesale cuts of meat and may not be visible on retail cuts.

FIGURE 15–2 Quality Grade Shields for Beef Used in Commercial Food Establishments

Best-quality grade

USDA
CHOICE

Second-best-quality grade

Third-best-quality grade

TABLE 15–1 Quality Grades of Meat (in Order of Best Quality)

BEEF	VEAL	LAMB	PORK
USDA Prime	USDA Prime	USDA Prime	USDA 1
USDA Choice	USDA Choice	USDA Choice	USDA 2
USDA Good	USDA Good	USDA Good	USDA 3
USDA Standard	USDA Standard	USDA Utility	
USDA Commercial	USDA Commercial	USDA Cull	
USDA Utility	USDA Cull		
USDA Cutter			
USDA Canner			

Grading for Yield

A **yield grade** is assigned to an animal carcass that has been graded according to the amount of lean, edible meat. The government has established five yield grades, which rate the yield of meat from 1 to 5. Yield grade 1 is the same as USDA Prime, and yield grade 5 is the same as USDA Commercial. Figure 15–3 illustrates a yield grade stamp for beef, pork, veal, or lamb. Each yield grade indicates a different proportion of lean, fat, and bone. The more salable the meat, the higher the carcass yield. The yield

FIGURE 15–3 Yield Grade Stamp

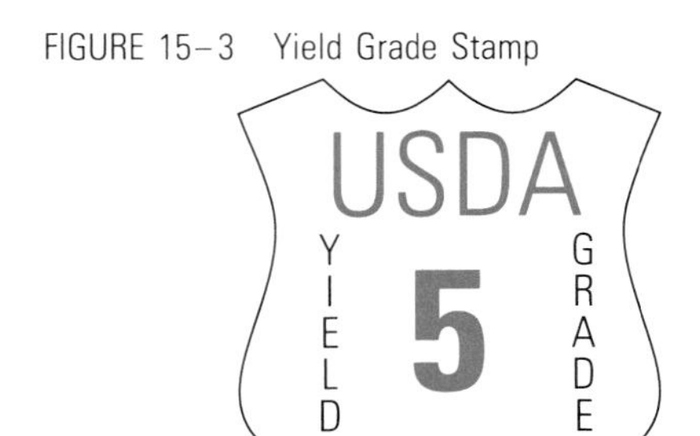

MATH SCREEN

Raw cuts of meat that have been trimmed yield approximately 75 percent edible meat after cooking. About 25 percent of their weight is lost in the cooking process. These losses primarily consist of water and fat drained off or evaporated during cooking, and they can be minimized by cooking meats at low temperatures.

Food service operators and cooks must be able to calculate these losses before purchasing to prevent running out of a product. Meat portions should be figured on the *edible portion* weight of a meat after cooking, not on the *as purchased* weight, which is often significantly different.

The following formula is used to calculate the percentage of weight lost during cooking:

AP = as purchased weight

EP = edible portion weight

$$\frac{AP - EP}{AP} = \text{percentage of weight lost}$$

Study this example: A cook has six pounds of eye of round roast. After cooking it at a low temperature (275°) to a medium stage, the roast weighs five pounds. What percentage of weight was lost during cooking?

$$\frac{AP - EP}{AP} = \text{percentage of weight lost}$$

$$\frac{6 - 5}{6} = .166$$ or 17 percent of weight lost (leaving 83 percent of the purchase weight as edible meat)

Now test your math skills on the following problem: Twelve pounds of rump roast were purchased and cooked at 375°F to a medium stage. After cooking, the weight of the roast was nine pounds. What percentage of weight was lost during cooking?

grade is an important factor to restaurant operators, especially those who run large food chains and prefer to purchase wholesale cuts and portion cuts according to their needs. These operators need to know how much meat they are getting for their money.

Types of Red Meat

Red meat is popular because it comes in so many varieties and can be prepared in countless ways. The main types of red meat are beef, veal, lamb, and pork.

Beef

Beef is the meat of steers, heifers, cows, bulls, and stags that are bred for human consumption. Quality beef is bright red in color with generous marbling in the lean part of the meat. The quality of beef is determined by the age and sex of the animal, the amount of marbling, and the color and texture of the meat. Beef is graded on its quality and yield.

ILLUSTRATION 15–1 Beef is the meat of steers, heifers, cows, bulls, and stags that are bred for human consumption.

Most beef sold as **retail,** the form sold in grocery stores, comes from animals between 15- to 30-months old and is stamped USDA Choice, USDA Good, and USDA Commercial. In commercial food service establishments, the most commonly used grades are USDA Prime, USDA Choice, and USDA Good.

Beef may be purchased by the whole carcass, side carcass, quarter carcass, wholesale cuts, primal cuts, retail cuts, and fabricated cuts. Fabricated cuts are ready to cook, pretimed, size portioned, and uniform. On the average, less than half of the weight of a fully grown animal is available for consumption after slaughtering. For example, a steer weighing 1,000 pounds will yield approximately 425 pounds of edible meat in the form of retail cuts.

Tender cuts of meat include the rib, short loin, and sirloin. They are best when broiled or roasted. Tougher cuts—the brisket, short plate, flank, and round, for example—are best when cooked with moist heat techniques, such as braising or stewing. Figure 15–4 (page 196) illustrates the retail cuts of beef, the most desirable methods of cooking them, and the location of the cuts in a carcass of beef.

Veal

Veal comes from milk-fed calves under twelve weeks of age. The quality of veal is determined by the age, color, and texture of the meat, and by the amount of fat that surrounds the

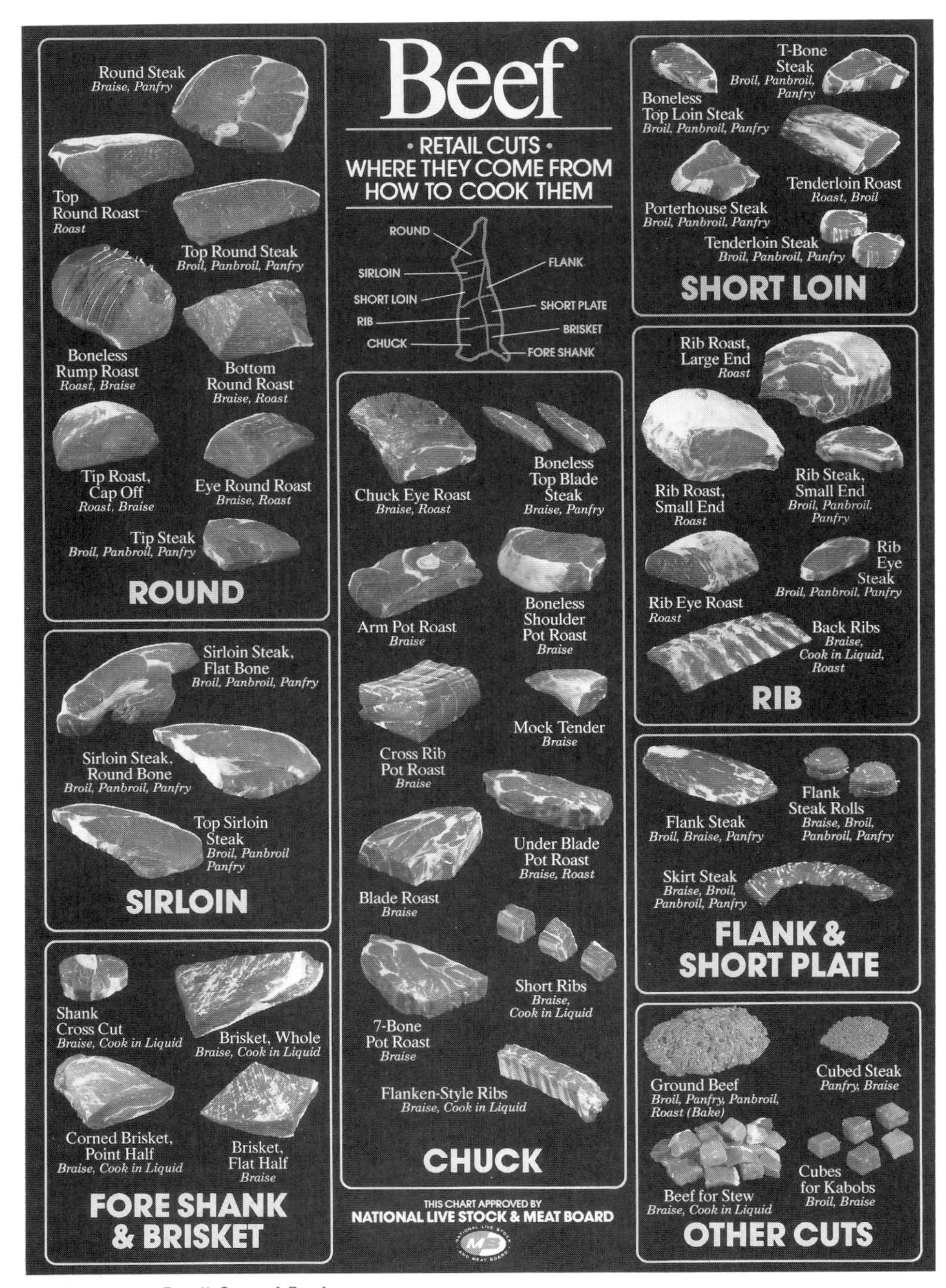

FIGURE 15–4 Retail Cuts of Beef

shoulders, rump, and kidney. Quality veal is light pink and has no marbling, a soft texture, and small bones. The best quality veal comes from calves slaughtered at less than eight weeks of age, when the carcass ranges between 50 and 250 pounds. Market forms available are the whole, side, and quarter carcass, as well as the wholesale, primal, fabricated, and retail cuts. Wholesale cuts are less expensive per pound than retail, but they require more labor hours for trimming and portioning.

Veal is a popular menu item in many fine restaurants. Its' nutritional value is superior to beef because, ounce-per-ounce, it contains less fat and cholesterol. Veal bones make excellent stocks and are the best type of bones to use when preparing soups and sauces.

Lamb

Lamb is the meat of sheep that are, generally, under 12 months of age. Roast leg of lamb and lamb chops are popular menu items in fine restaurants. Age and diet determine the quality of a lamb's meat. The best quality lamb comes from an animal that has been milk fed and slaughtered at less than 6 months of age. When lambs start to feed on grass, their meat becomes dark and tough. A lamb that has been milk fed has tender, light pink flesh and a generous white-fat covering around the kidneys. Lamb is marketed as a carcass (whole) or as a saddle (fore saddle and hind saddle). Wholesale, primal, fabricated, and retail cuts are also available.

Pork

Pork is the meat of hogs, including barrows, gilts, and young sows. Quality pork is grayish pink in color, soft in texture, and tender when cooked. Pork is an excellent source of thiamin, and it should be eaten weekly. In response to calorie- and cholesterol-conscious consumers, pork producers are breeding hogs that have less fat in their meat.

When pork is transported across state lines, it is carefully inspected for ***Trichinella spiralis,*** a microscopic worm that is found in pork products. The worm can be transmitted to humans, causing trichinosis, if the meat is not cooked to a well-done stage.

Pork may be purchased fresh, cured, or smoked, in the following marketed forms: picnic ham, head, Boston butt, belly, shoulder butt, spare ribs, hocks, and ham.

POULTRY

Poultry refers to birds that are slaughtered for human consumption. Poultry is generally classified according to age, because age is a major determinant for a bird's quality. After purchasing poultry, it should always be thoroughly washed, inside and out, to remove as much *Salmonella* bacteria as possible. *Salmonella* is commonly found in poultry products. After contact with poultry, all knives, cutting

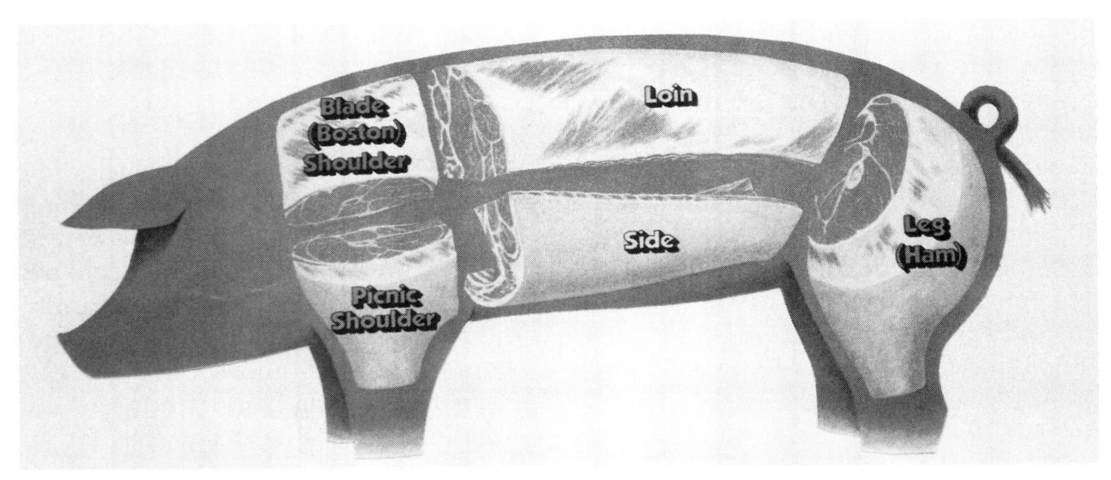

ILLUSTRATION 15–2 Pork is the meat of hogs, including barrows, gilts, and young sows.

boards, and counters must be cleaned and sanitized.

Inspection and Grading

Poultry is inspected for wholesomeness and quality by the USDA. All poultry grading is voluntary. When poultry is graded, it is judged on the basis of its shape, fat distribution, appearance, and the condition of its skin. United States poultry grades include USDA Grade A (this top grade is illustrated in Figure 15–5), USDA Grade B, and USDA Grade C. Poultry should be free of bruises, cuts, and pinfeathers. It should also have a good layer of fat under the skin.

Types of Poultry

The most popular types of poultry purchased in food service establishments are chicken, turkey, duck, pheasant, and quail. Goose is served mainly in fine restaurants. Poultry is versatile, and it can be purchased and prepared in many different forms.

Chicken

Chicken has tender white and dark meat. Its taste, flavor, nutritional value, and tenderness have made chicken a popular menu item at many food service establishments. In addition, chicken can be used to prepare a

FIGURE 15–5 USDA Grade Shield for Top-Quality Poultry

variety of menu entrées, including stocks, soups, sauces, salads, and appetizers.

Chicken is marketed in many ready-to-eat forms (meaning that all organ parts have been removed), such as roasters, fryers, broilers, capons, and cornish game hens. It can be purchased frozen, precooked, in rolls, barbecued, fried, stuffed, canned, and deboned. Chicken is sold whole, in halves, in quarters, and in small pieces. The most economical way to purchase chicken is to buy it whole and cut it into desired portions.

Turkey

Turkeys are marketed according to the age at which they are slaughtered; baby turkeys, young hens, old hens, young toms, and old toms are available. In addition, turkeys are bred as lightweight birds and heavyweight birds. Commercial food service establishments purchase only heavyweight turkeys because these birds have more meat in proportion to bone, and they usually sell at a lower price per pound. Baby turkeys (under 16 weeks) and young hens and toms (under 1 year old) have soft meat with flexible skin and breastbones. Young hens and toms average between 12 and 16 pounds, while baby turkeys average between 4 and 8 pounds. The lightweight breed of adult turkeys range between 12 and 16 pounds, and the heavyweights range between 18 and 30 pounds. Young turkeys can be roasted, sautéed, or fried. Older tur-

ILLUSTRATION 15–3 Poultry is inspected for wholesomeness and quality by the USDA.

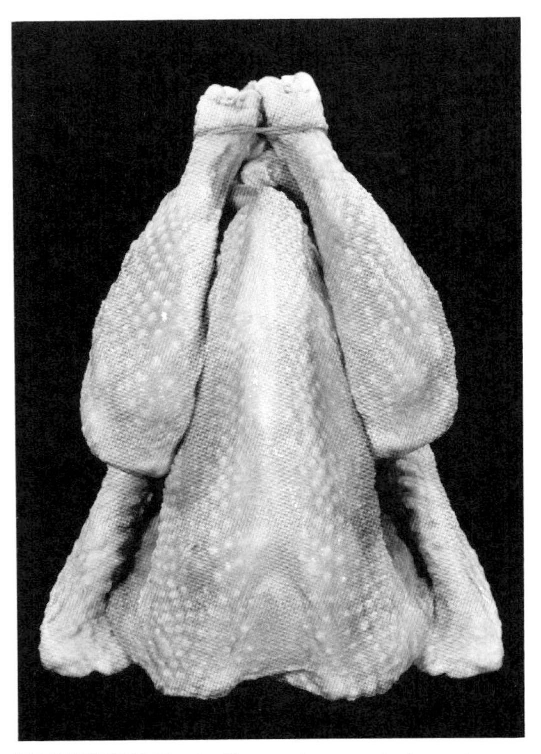

ILLUSTRATION 15–4 The most economical way to purchase chicken is to buy it whole and cut it into desired portions.

keys are usually roasted because they are less tender.

Duck

Ducks are often raised domestically. Some ducks, however, are *game birds,* meaning they have grown up in the wild. Ducks have dark meat only, and they have less flesh in proportion to bone and fat.

Young ducklings, which are under eight weeks old, have tender flesh and are sold as broilers or fryers. Ducklings are marketed according to their geographical breeding location. These birds weigh approximately 3 pounds. Young ducks—those just under 16 weeks of age—are sold as roasters and weigh approximately 4 pounds. Ducks over 6 months old are called *mature.* Their flesh is tough and can only be used in stews. They weigh approximately 4 pounds.

Pheasant and Quail

Pheasant and quail are both game birds and, as menu entrées, can only be found in expensive restaurants. Pheasant is a dark, flavorful meat. It is usually braised, roasted, or stuffed with wild rice. Quail is a white meat, and a single quail usually provides only enough meat for one person. Quail tastes best when it is sautéed or broiled.

FISH

There are more than two hundred species of fish consumed in the United States today. These species range from salt-water to fresh-water fish, from finfish to shellfish. Lean fish contains approximately 5 percent fat, and fat fish contains 20 percent or more fat. Fish with a high fat content have a different flavor than lean fish, and this fat content affects the cooking method. Trout, for example, is excellent when it is baked or broiled, because the fat prevents the fish from drying out. Lean fish is best when cooked with moist heat techniques, such as poaching or steaming. Unlike

ILLUSTRATION 15–5 There are more than two hundred species of fish consumed in the United States today.

red meat, finfish and shellfish have little connective tissue, and they can be cooked in a variety of ways without compromising their quality.

Inspection and Grading

The inspection and grading of fish is voluntary and is the responsibility of the National Marine Fisheries Service of the National Oceanic and Atmospheric Administration (which is part of the U.S. Department of Commerce). Fish is inspected on the basis of sanitary handling and storing procedures. The quality grades given to fish are the same as those given to poultry. Commercial food establishments use only U.S. Grade A fish (see Figure 15–6). The two lower grades are used in processed fish products.

Finfish and Shellfish

Finfish have bone structures and fins. They may be purchased fresh, frozen, cured, or smoked. Fish is best when it is fresh; it should be cooked shortly after it is removed from fresh or salty waters. When purchasing fish, food buyers should look for the following characteristics to ensure quality and freshness: bright eyes, bright red gills, firm flesh, shiny scales tightly adhered to the skin, and mild odor. In addition, fresh fish will sink when placed in water.

The most popular types of shellfish served in commercial food service establishments are shrimp, crab, lobster,

FIGURE 15–6 Grade Shield for Top-Quality Fish

NUTRITION NOTE THE IMPORTANT NUTRITIONAL CONTRIBUTION OF MEAT, POULTRY, AND FISH

Meat, poultry, and fish are rich in many nutrients, especially in high-quality protein and iron. The protein in meats is called complete protein because it contains all the amino acids that are essential for the growth, maintenance, and repair of body tissues. In addition, complete protein increases the body's ability to fight infection and disease.

The iron in meat, poultry, and fish helps red blood cells to carry oxygen from the lungs to the body tissues and to bring carbon dioxide back to the lungs. The iron found in red meats is called heme iron. This form of iron is easily absorbed by the body. An undefined factor in meat, fish, and poultry increases the absorption of nonheme iron from other sources (as does vitamin C). These foods are thus important components in a healthy diet.

oysters, scallops, and clams. Shellfish does not have bones; rather, it has an outer shell and a soft, tender, white meat with a distinct flavor, texture, and taste. Shellfish has minimal fat content and connective tissue. Market forms of shellfish include live, shucked (shell removed), headless, cooked, and canned. Shrimp, lobster, and oysters are graded according to the number per pound or gallon.

Finfish and shellfish are extremely perishable and require sanitary handling and appropriate holding and storing temperatures to prevent bacterial contamination. Freezer storage temperatures should be close to 0°F. Cooked fish can be held in freezer storage a maximum of three months. Raw fish should be stored in a freezer for six months or less. Frozen fish must be thawed in a refrigerator or under rapidly running cold water.

STOCKS, SOUPS, SAUCES, AND GRAVIES

Meat, poultry, and fish can be used to make a number of other food products. The most well-known of these products are stocks, soups, sauces, and gravies, which enhance the flavor of other foods and, in the case of soups, can be served as entrées.

Stocks

Stock is a liquid ingredient produced by simmering meat, fish, or poultry bones and scraps for several hours in water until their flavor and nutrients have been extracted and the water has become concentrated. The best stock is made from young animals, such as calves or lambs, because they have more cartilage than older animals. Cartilage and collagen melt during the cooking process and form a gelatinous substance that enhances the richness and flavor of a stock. A stock's richness depends on the proportion of bones to water during the cooking process.

Stock is the base ingredient used to prepare soups, sauces, and gravies. Stock enhances the flavor of foods and adds to their nutritional value.

Soups

Soup is a liquid food prepared from the stock of meat, poultry, fish, shellfish, or vegetables. It can be served as an appetizer or as an entrée. Appetizer soups are lightly flavored with few food particles, while entrée soups have hearty flavors and thick consistencies filled with food particles.

Thick Soups

Thick soups have a viscous consistency. They may be thickened with flour, cereals, potatoes, pasta, or vegetables. (The use of flour as a thickening agent is discussed in more detail in Chapter 17.) Thick soups are considered hearty soups because they are very filling.

Clear Soups

Clear soups include broth (or bouillon), consommé, clear vegetable, and borscht made from meat, chicken, fish, or vegetable stock.

Bouillon is a French word for broth that is made from meat, poultry, or fish stock. All visible fat is skimmed off. Popular variations include English beef broth with barley and potatoes, lamb broth with barley and tomato puree, and chicken broth with vegetables.

Consommé is a concentrated soup made from the clear broth of meat, poultry, or fish, plus other ingredients such as lean ground beef, ground vegetables, egg whites, and ice water. After all the ingredients are combined, the soup is put through a clarification process. **Clarification** is the removal of sediments. This process produces a clear soup that can be decanted.

Vegetable soups are made with poultry, meat, or fish bouillon and a variety of vegetables. **Borscht** is a beet soup with diced onions, sugar, and vinegar or lemon juice. It can be served hot or cold, often with sour cream.

Cold Soups

Cold soups are served at refrigerator temperatures, between 40°F and 45°F. Popular cold soups include jellied consommé, vichyssoise, gazpacho, and borscht. Jellied consommé is a tomato soup made with white stock and served in a chilled bowl with a slice of lemon. **Vichyssoise** is a potato and leek soup made with chicken stock and mixed with chilled cream. **Gazpacho** is a highly seasoned tomato and vegetable soup made from beef stock and served with finely chopped cucumbers and tomatoes.

Sauces

A **sauce** is a thickened, flavorful liquid that enhances the taste of prepared foods. Sauces are classified as warm, cold, or sweet. The basic ingredients needed to prepare a variety of sauces are stock, milk (or cream), butter, eggs, sugar, and fruit.

Warm Sauces

Warm sauces are classified into five categories from which hundreds of other sauces can be made. The five categories of warm sauces are brown or *espagnole,* cream or béchamel, Ue-loaté or fricassee, hollandaise, and tomato. These sauces may be thickened with roux, slurry, or cornstarch. Roux is made with equal parts of flour and water. Brown roux is used to thicken dark sauces, and it is made by cooking white roux until it becomes brown. If cornstarch is used as a thickener, it should first be mixed with cold water, cold milk, or cold stock, and then added to the hot liquid. This method prevents lumping.

A good warm sauce should have a flowing, lump-free consistency, a smooth and velvety feel, a delicate flavor without a starchy aftertaste, and a rich color with a slight sheen appropriate for the type of sauce being made.

Cold Sauces

Cold sauces, also called dressings, may be prepared from a variety of different foods. The most popular cold sauce is mayonnaise. Cold sauces may be sweetened as in sweet sauces, or they may be unsweetened as in salad dressings or mayonnaise.

Sweet Sauces

Sweet sauces are made with sugar, fruit, and fruit juices in a variety of combinations. They may also be prepared with milk, cream, or chocolate. Sweet sauces are used mainly as dessert fillings or toppings, but they are also popular with lamb, duck, and ham entrées. Sweet sauces are thickened with caramelized sugar. Their consistency ranges from thin to thick, depending on the type of dessert or entrée. Sweet sauces may be served warm or cold.

Gravies

Gravies are made from meat or poultry drippings and juices that have concentrated during roasting. Gravies are thickened with white or brown roux and are usually seasoned with salt and pepper. White gravies, such as for chicken and turkey, may be thickened with white roux and cream for an extra rich flavor. As when making warm sauces, the roux must be mixed with cold water before being added to the gravy; otherwise, the roux will cause lumps. Spices and herbs can be used for lamb and pork gravies to enhance their flavor, but they must be used in moderation so that they do not dominate the gravy's natural flavor.

CAREER PROFILE
POSITION: NUTRITIONIST FOR A LARGE SCHOOL DISTRICT

Qualifications:

- Bachelor's degree in nutrition, dietetics or food service management
- Experience in quantity food preparation, recipe standardization, and menu planning is desirable
- Any equivalent combination of training, education, or experience that meets the minimum qualifications

Responsibilities:

- Plans menus and ensures that they comply with established federal regulations for nutrient content and menu pattern
- Determines food products to be used in menu items (often choosing between government commodities and vendor-purchased items)
- Surveys students and conducts taste tests for potential products
- Prepares the portion menu for each site to ensure compliance with government regulations regarding the quantity of food to serve
- Prepares worksheets on projected menu needs to inform the kitchen staff of the number of food components that must be met by an entrée item
- Ensures that products for the menu items are available as needed
- Prepares bid specifications for food and paper products and ensures that the specifications for food meet the USDA requirements for nutritional value
- Prepares and ensures that standardized recipes are used in the preparation of all food items
- Recommends and coordinates the implementation of new programs in the schools and investigates parent complaints
- Assists in planning nutrition education programs offered in the schools and conducts nutrition education lessons

SUMMARY

Meat, poultry, and fish are important components of a healthy diet, and they make up a large portion of the food cost for restaurants and other food service establishments. A dizzying variety of meat, poultry, and fish is available. These foods are the key ingredients in many popular entrées, from beef Wellington to pork chops, from roast turkey to stuffed pheasant, from smoked salmon to lobster tails. Because these items are so important to food service businesses, the careful selection and purchasing of meat, poultry, and fish helps to minimize waste and cut food and labor costs.

The inspection and grading of meat, poultry, and fish products is done by the U.S. Department of Agriculture and the National Marine Fisheries Service. These agencies ensure that animals, birds, and fish are slaughtered and processed under sanitary conditions. They also rate the quality of these foods, stamping quality grades or wholesale carcasses or packages.

A variety of stocks, soups, sauces and gravies are prepared from the bones and scraps of meat, fish, and poultry. These items add variety to menus and are a way of providing numerous nutrients in one serving of food.

QUESTIONS

1. What are three factors that contribute to the marbling of red meats?
2. What is meant by *wholesomeness?*
3. What is quality grading for meats based on?
4. What are the eight quality grades given to beef products? List the grades in order from highest to lowest.
5. What is the difference between wholesale and retail cuts?
6. What does a yield grade signify?
7. What are the most tender cuts of beef? In what part of an animal carcass are they located?
8. What is the reason that pork products must be cooked to a well-done stage?
9. What U.S. Grade of fish products is used in commercial food establishments?
10. It is important to include pork in the diet at least once a week because it is rich in what vitamin?
11. Meat, poultry, and fish are the primary source of what nutrient?
12. In what ways are stocks, soups, sauces, and gravies different, and in what ways are they similar?

ACTIVITIES

Activity One Purchase the ingredients necessary to prepare the brown stock and the brown sauce recipes at the end of this chapter. Prepare the stock, and then cool and store it in the refrigerator until ready for Activity Two.

Activity Two Using the stock from Activity One, prepare the brown sauce recipe. Evaluate the sauce on the basis of taste, flavor, consistency, and color.

RECIPES

Brown Stock

Prepared Ingredients	Weight	Measure	Method
Veal bones, cut in large portions	4 lbs		Place bones in a roasting pan and brown in a 400°F oven. Drain fat. Add vegetables and seasoning and return to the oven until vegetables are slightly brown. Transfer bones, vegetables, and seasoning into a large sauce pot. Deglaze the roasting pan with two cups
Lamb bones, cut in large portions	4 lbs		
Water		2 gal	

Vegetables (*mirepoix*):			of the water until all brown particles at the bottom of the pan have been removed. Pour the liquid into the sauce pot with the bones. Add the remaining water and cook uncovered for five to six hours until half of the water has evaporated. Strain through a fine china cap into a two gallon container. Cool as quickly as possible by placing stock container in a large container with ice and stirring quickly. Refrigerate until ready to use.
Celery, coarsely chopped	¼ lb		
Carrots, coarsely chopped	¼ lb		
Seasonings (*bouquet garni*):			
Bay leaves		2	
Thyme		½ tsp	
Cloves		2	
Garlic		2	
Black pepper to taste			
Tomato puree		12 oz	

Quantity: Makes 1 gallon of stock (to be used as a base for other recipes).

Serving Suggestion: May be used to make brown sauce for gravy.

Brown Sauce

Number of Servings: 24

Prepared Ingredients	**Weight**	**Measure**	**Method**
Brown stock		½ gal	In a large sauce pot bring stock to a simmer. Add vegetables and seasoning ingredients and continue to simmer.
Celery, coarsely chopped	¼ lb		
Carrots, coarsely chopped	¼ lb		
Bay leaf		1	Meanwhile, melt butter or shortening in a sauce pot over medium heat. Add flour and cook for five minutes or longer until it browns. Add this hot brown roux into the hot stock pot and stir until roux is completely dissolved and sauce thickens. Continue to simmer for two hours, stirring frequently. Strain sauce through a china cap into a clean container while forcing the vegetables through the china cap sieve. Stir sauce. Serve over your favorite meat while hot.
Thyme		½ tsp	
Salt and pepper to taste			
Roux:			
All-purpose flour	¼ lb		
Butter or shortening	¼ lb		

Serving Size: 2 ounces.

Serving Suggestion: May be served over a roast leg of lamb.

16 VEGETABLES AND FRUITS

VOCABULARY

pigments
anthocyanins
anthoxanthins
carotenoids
chlorophyll
pH
pheophytin
sodium
milligram
fructose
scurvy
enzymatic browning

OBJECTIVES

After studying this chapter you should be able to do the following:

- Explain the changes that occur in the pigment of fruits and vegetables during cooking
- Identify the different vegetable and fruit classifications
- Explain why the proper selection and storage of fruits and vegetables is important
- Identify the important nutrients found in fruits and vegetables
- Describe the major components of a salad

Vegetables and fruits are important components of every menu. In addition to the valuable nutrients they contain, fruits and vegetables add color, texture, and flavor to a meal. They can be purchased fresh, frozen, canned, dried, or juiced, and they may be used as an entrée, accompaniment, appetizer, salad, or beverage. Some vegetables are used as herbs and spices to season a wide variety of dishes. Selecting and storing fruits and vegetables according to their different forms helps to preserve their quality and nutritional value as well as their flavor.

THE COMPOSITION AND CHARACTERISTICS OF VEGETABLES

Vegetables consist of water, starch, and sugar. They are low in calories and a major source of minerals, protein, and vitamins A and C (as well as some B vitamins). Vegetables come in four natural types of **pigments,** or coloring substances: **anthocyanins**—red and purple pigments; **anthoxanthins**—colorless or white pigments; **carotenoids**—orange, yellow, red-orange, and red pigments; and **chlorophyll**—green pigment. All these pigments are affected by acidic or alkaline mediums. A **pH** level is the measure used to indicate the acidity or alkalinity of an object or solution. Acidic mediums have a pH less than seven, and alkaline mediums have a pH greater than seven. A pH of seven is neutral. Changing a fruit or vegetable's pH level affects its color. Table 16–1 illustrates the ways that vegetable pigments change when they are treated with acidic or alkaline solutions.

Vegetables and fruits that contain anthocyanin pigments include red cabbage, beets, cherries, and plums. To avoid pigment changes, these vegetables and fruits should be cooked by waterless methods, such as steaming. Water increases in alkalinity during cooking, so, if water is used, a cook should add vinegar or lemon juice (both acids) to offset this process.

Potatoes, onions, cauliflower, and turnips contain anthoxanthin pigments. These vegetables should be cooked with very little water to prevent color changes. Large amounts of highly alkaline water cause vegetables with anthoxanthin pigments to develop a yellowish tint. If this happens, the addition of small amounts of lemon juice will restore the original color.

Vegetables with carotenoid pigments, such as carrots and squash, do not change colors during cooking. Chlorophyll, however, the pigment responsible for the rich color of all green vegetables, is affected by cook-

TABLE 16–1 pH Influence on Pigmentation

PIGMENT	NATURAL COLOR	COLOR WHEN TREATED WITH ACIDIC SOLUTION	COLOR WHEN TREATED WITH ALKALINE SOLUTION
Anthocyanin	Red	Red	Blue
	Purple	Red	Blue-green
Anthoxanthin	Colorless	White	Creamy white
	White	White	Yellow
Carotenoid	Orange	Natural color	Natural color
	Yellow	Natural color	Natural color
	Red	Natural color	Slightly brown tint
Chlorophyll	Green	Brownish green or yellow-green	Bright green

ing. Green vegetables include broccoli, beans, and peas. Chlorophyll retains its rich green color in alkaline solutions but turns yellow or brownish green in acidic solutions. Excessive heat is another problem. It causes the formation of **pheophytin,** a substance developed when the acids in green vegetables remove a magnesium ion from chlorophyll, causing the vegetables to change color from bright green to olive green or grayish green. To prevent these undesirable color changes, all green vegetables should be cooked in tap water that is slightly alkaline. In addition, acids from the vegetables should be allowed to escape during cooking by uncovering the pot and stirring the vegetables as soon as they start boiling.

TYPES OF VEGETABLES

Vegetables are classified as flower, fruit, green leafy, seed, root, or stem vegetables. The part that is eaten varies with each vegetable. The following sections describe the selection, use, and preparation of these different types of vegetables.

Flower Vegetables

The most familiar flower vegetables are broccoli and cauliflower. They can be served raw (as appetizers), in salads, or cooked.

Broccoli

Purchased broccoli should have green heads, leaves, and stems. The stalks should be tender and firm, and the cluster of buds on the top of each head should be closed. When preparing broccoli, the stems should be slashed lengthwise and cooked with the heads (which are made up of florets). These techniques ensure that the cooking times for both the stems and the florets are the same.

Cauliflower

Quality cauliflower is white or creamy white. The head should feel heavy and compact, and the leaves surrounding the head should be fresh and crisp. Size is not an indicator of quality. The most popular way to serve cooked cauliflower is to top it with cheese sauce, lemon juice, or tartar sauce. Cauliflower may also be sautéed, fried, or added to stews.

Fruit Vegetables

Fruit vegetables include avocados, cucumbers, eggplants, peppers, squash, and tomatoes. These vegetables (except eggplants) are often eaten raw.

Avocados

Avocados are considered fruits and vegetables, but they are most closely allied to vegetables. Their flavor is rich and buttery, and they are usually served with lemon or lime juice.

Cucumbers

Cucumbers average from 10 to 12 inches in length with some varieties up to 3 feet long. Cucumbers are more than 95 percent water. They are

ILLUSTRATION 16-1 Vegetables are classified as flower, fruit, green leafy, seed, root, or stem vegetables.

cool, crisp, and low in calories. Cucumbers are excellent when served with a dip or added to a salad.

Eggplants

The eggplant is a purple-black, glossy, firm vegetable ranging from 10 to 12 inches long. Smaller varieties are generally better in quality. They vary from black to purple, green, yellow, or white. Eggplant is served cooked only.

Peppers

Peppers come in different varieties and colors. Sweet peppers of good quality have a thick, dark green flesh. A firm pepper may crack with slight pressure. Chili peppers can be cone shaped or button shaped. Peppers are an excellent source of vitamin C. They can be served raw, as a seasoning for other dishes, or as a main entrée.

Squashes

Squashes come in a variety of shapes, sizes, and colors. Popular varieties include yellow straightneck, yellow crookneck, and zucchini. Quality squashes have firm, shiny surfaces and are free from bruises. In general, the smaller the squash, the more tender the flesh, seeds, and skin. Squashes can be baked, broiled, creamed, steamed, sautéed, or fried. Zucchini is the most popular variety of squash. It is usually eaten raw with a dip or added to a salad.

Tomatoes

The best-tasting tomatoes are those allowed to ripen on the vine. Quality tomatoes are plump, well formed, smooth, and reddish in color.

Tomatoes are a delicious addition to many dishes, and they make a great salad with lettuce, cucumber, and onions. They are also tasty when served sliced with no dressings or seasonings. Tomatoes may be stewed, fried, baked, pickled, boiled, or juiced. A tomato's size determines its use. Large green tomatoes are excellent for frying. Jumbo red tomatoes are good plain, stuffed, or baked. Medium-sized tomatoes are good for slicing, and small tomatoes are the best for pickling. Overripe tomatoes work well in stews, sauces, and casseroles.

Green Leafy Vegetables

Green leafy vegetables include beet greens, turnip greens, mustard greens, kale, collard, spinach, dandelion, endive (also called escarole), chard, parsley, mint, and lettuce. The most common varieties of lettuce include butter head, bibb, iceberg, red leaf, and romaine.

Compared to other vegetables, greens are high in vitamins A and C, iron, and magnesium. Green leafy vegetables also add color and variety to meals at easily affordable prices.

Quality greens are crisp, young, undamaged, and have a healthy green color. Each type is best when purchased at the height of its season.

With the exception of lettuce, mint, and parsley, green leafy vegetables can be purchased fresh, frozen, or canned. Mint and parsley may also be purchased in a dried form.

Seed Vegetables

Beans, peas, and corn are seed vegetables. These vegetables are popular items in many food service facilities.

Beans

The most common types of beans are snap, lima, soy, and wax beans. Beans are either green podded, yellow podded, or wax, and their shape ranges from flat to round. The quality characteristics to look for in snap beans include a fresh, clean appearance and a tender, crisp texture without cracks, bulges, or scars. Snap beans can be purchased fresh, frozen, and canned, and they may be cooked in a variety of ways.

Peas

Peas come as chick-peas (or garbanzos), cowpeas (or black-eyed peas), green peas, and snow peas. Peas should be picked as soon as their pods are well filled, and they should be cooked or processed immediately after harvest to preserve their vitamin and sugar content. Peas can be purchased

NUTRITION NOTE HERBS AND SPICES FOR LOW-SODIUM DIETS

Sodium is a chemical element that is needed to maintain proper water balance in the body. Certain health conditions, however, cause the body to retain sodium, which in turn forces the body to hold more water. A condition like hypertension (also called high blood pressure) for example, causes water retention, swelling, and discomfort. Decreasing sodium intake helps the body to lose this extra water. Salt does not cause the disease, but it further aggravates the condition.

Sodium is plentiful in all foods and is the main component of table salt. Table salt is approximately 40 percent sodium. One teaspoon of table salt contains 2,200 milligrams of sodium. A **milligram** (mg) is equal to 1 one-thousandth of a gram. The recommended dietary allowance (RDA) for salt is 1,100 to 3,300 milligrams, which is equal to ½ to 1-½ teaspoon of salt daily. The average American consumes from 2 to 4 teaspoons of table salt each day.

Because of the risks involved with the overconsumption of sodium, millions of Americans are on low-sodium diets. Since this type of diet has become so prevalent, it is important to seek alternative methods to flavor foods.

Herbs and spices can provide a creative and tasteful alternative to salt. The skillful use of herbs and spices can turn a simple meal into a gourmet dish. In addition, spices and herbs play a valuable role in good nutrition by helping to increase the appeal and appreciation of foods that are healthful. Except in some restrictive diets, herbs and spices can be used to help people stay faithful to an otherwise distasteful diet regime.

All herbs and spices are low in sodium and calories (with the exception of dehydrated celery and parsley flakes, which are slightly higher in sodium content). There are no strict rules for using herbs and spices. A good rule, however, is not to use two strong flavors together, but rather to use one strong and one mild flavor. Dried herbs and spices are stronger than fresh ones, and powdered herbs are stronger than crushed ones.

The following rule is useful for measuring herbs and spices: ¼ teaspoon powdered herbs = ¾ crushed = 2 teaspoons fresh. Leaves should be chopped finely for more surface exposure and thus more flavor absorption. The list that follows shows recommended measurements for different seasonings:

- Strong herbs and spices—use approximately 1 teaspoon for 6 servings. These include bay, cardamom, curry, ginger, hot pepper, sage, and mustard.
- Medium herbs and spices—use 1 to 2 teaspoons for 6 servings. These include basil, celery seed, dill, fennel, mint, oregano, and tarragon.
- Delicate herbs and spices—use in generous amounts or combine with medium and strong spices. These include chives and parsley.

fresh, frozen, or canned. For the best flavor and nutritional value, peas should be cooked with steam.

Corn

Quality corn has solid, evenly filled ears. For the best flavor and texture, it should be cooked soon after picking. Corn can be boiled or grilled while on the cob, and it can be cooked in a variety of ways as kernels only. The ever-popular popcorn is a variety of corn grown especially for its small ears and pointed kernels that explode when heated.

Root Vegetables

Americans eat a number of root vegetables. Beets, carrots, radishes, onions, and potatoes are the most popular.

Beets

Beets were originally grown for their leaves and not their roots. Today the roots are by far the most popular part. Quality beets are free from cracks and blemishes. Smaller beets are preferred to larger ones mainly for appearance, but also because they can be cooked whole in shorter periods of time. Beets can be purchased fresh, frozen, or canned. Canned beets come sliced, diced, quartered, julienned, and whole. They are a popular choice in salad bars.

Carrots

Carrots should be firm, clean, well shaped, and free from spots. Quality carrots have long rootlets and a bright orange color. Carrots contain a large amount of carotene, an orange pigment that is convertible to vitamin A. One medium-size carrot, for example, provides approximately 150 percent of the recommended dietary allowance (RDA) for vitamin A. Carrots can be served raw as a snack, or they can be cooked to accompany a main dish. They can also be used as a garnish or in salads, soups, and desserts.

Radishes

Radishes are root vegetables from the mustard family. They are also cousins of the cabbage. Their roots are firm and crisp, and their flavor is mild. The bright red color of radishes makes them popular as garnishes and appetizers.

Onions

The taste of onions is popular throughout the world. Onions should have bright, well-shaped bulbs, thin necks, and blemish-free skin. Large onions are best when sliced or chopped for use in stuffing or casseroles. Medium or small onions are excellent for boiling or cooking with roasts or stews. Red onions are mostly used in salads, and they make an attractive garnish for many dishes. Any type of onion that is pulled before maturity is called a *green onion,* or a *scallion.* Chives are miniature green onions. Leeks, which are giant green onions, are more mild in flavor than other onions.

Potatoes

Potatoes are the world's most popular vegetable. They are served daily as part of a meal in over 60 percent of American households. The most common forms of potatoes are the long russet, long white, round russet, round white, and round red. Quality potatoes have a smooth, firm skin that is free from cuts, bruises, or any green color. Well-shaped potatoes prevent waste during preparation.

Sweet Potatoes

The sweet potato is a native American vegetable. There are two types of sweet potatoes: light orange and dark red. Dark red sweet potatoes have a higher moisture content when cooked, and they are much softer than the light orange variety. One medium sweet potato, boiled and peeled, provides more than twice the RDA of vitamin A for an adult. In addition, it contains vitamin C, iron, and thiamin. Quality sweet potatoes are clean, smooth, well shaped, and firm.

Stem Vegetables

Asparagus and celery, which originated in Europe, as well as mushrooms are all classified as stem vegetables. These vegetables can be served raw or cooked in a variety of ways.

Asparagus

Quality fresh asparagus is firm and tender with closed tips and sturdy stalks. Canned and frozen asparagus are marketed as stalks, tips, or cut. When preparing asparagus, parts of the stalks that are too hard can be easily snapped off at the weak point separating the tender from the brittle sections. Asparagus is usually steamed while standing upright in a pot, with the tips on top.

Celery

Celery is a leafy, stalked herb. Quality celery is light green with a crisp and tender texture. Celery is a familiar ingredient in many soups and stews. It can be eaten raw as an appetizer and is often accompanied by a dip.

Mushrooms

Mushrooms have recently gained popularity and are available year-round in fresh, frozen, canned, or dehydrated forms. Quality mushrooms are white with a clean, smooth texture. They can be served raw, as appetizers or in salads, and they can be deep-fried or cooked in a variety of sauces, soups, stews, and stir-fry recipes.

TYPES OF FRUITS

Fruits are one of nature's most delicious and refreshing foods. They are an excellent source of dietary fiber and many important nutrients. In addition, most fruits are low in calories. They can be used as appetizers or in a range of salads, soups, main entrées, desserts, and snacks. The sweetness of fruit comes from **fructose,** which is a natural form of sugar. Fruits may be purchased fresh, frozen, canned, dried, or as preserves. They are classified as summer, winter, or tropical.

Summer Fruits

Summer fruits are popular because of their sweet taste and juicy textures. Most summer fruits are delicious raw, and they can also be baked or cooked in many different foods.

Berries

Berry varieties include blueberries, raspberries, boysenberries, strawberries, and blackberries. Berries are very tender. They should be picked on a cool morning, handled very little, and processed or served as soon as possible for best quality. If picked berries cannot be used quickly, they should be stored in a refrigerator until used.

Cherries

Cherries come in different varieties—some are sour and some are sweet. The most popular varieties are black. Light red cherries have a tart flavor and are best when used as maraschino or candied cherries, or when canned for baking. Fresh cherries should be refrigerated in a plastic bag and washed just before use.

Figs

Figs are green or black. They are best when eaten raw, with or without the skin. Quality figs have a velvety, ridged skin with some wrinkles on the surface. The inside of the fruit should be delicate and spongy. Figs can also be purchased candied or dried.

ILLUSTRATION 16-2 Fruits are one of nature's most delicious and refreshing foods.

Grapes

Grapes are fully ripe when they become sweet and aromatic. Unlike most other fruits, grapes do not improve their aroma after they are picked. Grapes are best when washed and eaten raw. They are also good, however, when dried as raisins.

Melons

Melons come in many varieties. The most popular are watermelon, cantaloupe, honeydew, crenshaw, and Persian. Melons should be sweet and juicy with a green, deep red, or yellow-orange color, depending on the type of melon. The best indicators of quality are an appropriate weight and slightly soft ends. Melons are low in calories and high in nutritional value. A 4-ounce slice of cantaloupe, for example, has only 24 calories and is a good source of many vitamins and minerals. Melons can be served as appetizers, salads, or desserts. They can be cut into many decorative shapes, such as wedges, balls, or slices, for use in fruit trays and salads.

Peaches and Nectarines

Peaches and nectarines are delicious and nutritious fruits that are low in calories and rich in fiber. Both these fruits are handpicked when firm and mature. Storing peaches and nectarines in a refrigerator slows down their ripening processes. These fruits are most flavorful, however, when served at room temperature. Peaches and nectarines may be purchased fresh, frozen, canned, dried, preserved, and juiced.

Plums

Quality plums are firm and plump, and they can ripen at room temperature within a few days. Some plum varieties soften and ripen rapidly, especially if exposed to high temperatures. Plums are nutritious and flavorful, and they add color and texture to many dishes.

Pears

Pears do not ripen on the tree and are usually harvested green. Pears are held in cold storage to stimulate ripening. When fully matured, bartletts are yellow in color and give off a sweet fragrance. Quality pears should be free from cuts and bruises.

Winter Fruits

Citrus fruits and apples are considered winter fruits. These fruits are nutritious and colorful.

Citrus

Citrus fruits include oranges, grapefruits, tangerines, tangelos, kumquats, lemons, and limes. When citrus fruits are scarce, people are often stricken with **scurvy,** an illness that weakens and causes pain in the muscles and joints. Scurvy is caused by a deficiency of vitamin C, which is abundant in citrus fruits.

Quality citrus is heavy for its size. The heavier the citrus, the juicier it will be. A rough skin texture may indicate that the fruit has a thick skin and a small amount of flesh. Dull, dry skin with a spongy texture is a sign that the fruit is old and probably of poor quality. Citrus fruits may be served fresh, candied, and juiced; as appetizers, garnishes, and desserts; or as part of salads or main dishes.

Apples

Quality apples are firm, bright, clean, and well colored. Underripe apples can be placed in storage at 60°F and allowed to ripen before use. Apples should be stored in a refrigerator and wrapped individually with paper to retain their flavor and prevent spoilage.

Apples are either tart or sweet. Tart apples are used mostly for cooking or baking pies, main dishes, beverages, and desserts. They can also be baked or fried to accompany a main course. Sweet apples are eaten raw.

Tropical Fruits

The variety of tropical fruits available to consumers has increased in recent years. The following sections discuss several of these fruits.

Bananas

Bananas are grown extensively in all tropical countries. This fruit varies in length from four to twelve inches, and from one-fourth to one inch in diameter. Bananas are rich in carbohydrates, fiber, vitamins, and minerals. Fully ripe bananas are brown flecked, all brown, or red, depending on the variety. They taste great raw. Green, unripe bananas should be placed in a cool, shady place to develop their flavor and sweetness. Green bananas can be used for cooking.

Kiwifruits

Kiwifruits are approximately the size of lemons. They have a fuzzy skin and a two-tone green flesh. Ripe kiwis are soft and juicy with a slightly sour taste. Kiwifruits make an excellent garnish or a colorful addition to fruit salads.

Mangos

Mangos are medium-sized, thick-skinned fruits approximately two to four inches wide and three to seven inches long. The outer skin of ripe mangos is either purplish red in color or yellow with red spots, depending on the variety. The flesh is light yellow to orange. Quality mangos are juicy and have a good flavor and texture. They make an excellent addition to fruit salads.

Papayas

Ripe papayas are greenish orange with a thin outer skin. They have a soft, juicy, pink-orange flesh and a central mass of black seeds. Papayas are an excellent source of vitamins A and C. Their vitamin C content increases as they ripen. Ripe papayas may be eaten raw, and unripened papayas can be cooked and served like vegetables. They are almost never frozen because freezing destroys some of their flavor and texture.

Pineapples

Fresh, ripe pineapples are firm and clean without visible signs of decay. Pineapples can be purchased fresh, and their outer shells can be used as decorations. Canned pineapples are available in sliced, chunk, or crushed forms, packed in their own juices or in a heavy syrup.

GRADING FOR CANNED FRUITS AND VEGETABLES

Canned products rated U.S. Grade A Fancy are the best that can be purchased. Their colors and flavors are excellent, and the sizes and shapes of the vegetables and fruits are perfectly uniform.

Canned products rated U.S. Grade B Choice are second best in quality and have good colors and flavors. Their individual pieces may vary in size and shape, with slight imperfections.

A U.S. Grade C Standard rating is given to poor quality products, which have some imperfections, odd shapes, and bruises. These products may be soft and mushy.

GRADING FOR FRESH FRUITS AND VEGETABLES

Grades for fresh fruits have been established and used mostly in wholesale trading. Grades are given on the basis of maturity, decay, waste, appearance, and shipping quality. These grades include U.S. Extra Fancy, U.S. Fancy, U.S. No. 1, U.S. No. 2, and U.S. No. 3. Fruits, however, are graded before shipping; temperature and handling during shipping can affect the quality upon receipt.

Because fresh vegetable grading is voluntary, few vegetables carry grades at the retail level. The major trading grade is U.S. No. 1, which represents average quality.

FRUITS AND VEGETABLES USED IN SALADS

Salads are popular, healthful, and profitable menu items that can be made with a wide variety of vegetables, fruits, cold meats, chicken, and fish. When preparing fruit and vegetable salads, appearance is very important. All the ingredients in the body of a salad should be bite-size, and they should represent a mixture of different textures. Different textures make a salad more pleasing. Care must therefore be taken not to bruise or wilt the lettuce when mixing. Green romaine lettuce leaves, red ripe toma-

ILLUSTRATION 16-3 Salads are popular, healthful, and profitable menu items that can be made with a wide variety of vegetables, fruits, cold meats, chicken, and fish.

toes, yellow peppers, and sliced red onion rings arranged artistically on an appropriate plate is one example of a simple, colorful, appetizing salad.

Salads are generally served on well-chilled plates. Often, they are named after the main salad component. All salads should include the following items:

- Base—usually a bed of leaf lettuce
- Body—the main component of the salad
- Dressing—a variety of different flavors is available, including blue cheese, ranch, Thousand Island, French, and Italian
- Garnish—may be a radish rose, a lemon wedge, a carrot stick, or an olive

Salad bars are well-accepted in popular restaurant self-service units. They include a variety of different salad bases, bodies, garnishes, and dressings.

STORAGE AND PREPARATION OF FRESH FRUITS AND VEGETABLES

Fresh fruits should be stored in a refrigerator with a relative humidity around 90 percent and a temperature between 36°F and 45°F. Fresh fruits and some vegetables continue to ripen while in storage. Storing them in a humid environment prevents the moisture losses that usually occur during ripening. The nutritional content of vegetables, as well as their texture, flavor, and appearance, diminishes when stored for long periods. Daily purchasing of fruits and vegetables ensures the best possible quality and freshness. Potatoes, onions, and garlic may be stored at room temperature longer than other vegetables; however, they must be stored in an area that has good ventilation and a temperature below 65°F.

Most fruits and vegetables are washed and cut just before storage. Some fruits, such as those in the melon and citrus families, need to be stored in sealed containers after cutting to prevent odors from being absorbed by other foods. Fruits such as bananas, nectarines, pears, and apples should only be cut just before serving to prevent enzymatic browning. **Enzymatic browning** occurs when the enzymes in a piece of cut fruit react with oxygen, causing the fruit's flesh to turn brown. If fruits prone to enzymatic browning need pre-preparation, they must be dipped in a water and acid solution to prevent browning. Lemon juice and water or pineapple juice and water are commonly used acidic solutions.

Delicate fruits, such as strawberries, should be washed and sliced just before serving to maintain their texture and firmness. Slicing berries before storage causes these fruits to lose some of their water content and become soft and mushy. Also, exposing cut fruits and vegetables to air results in some vitamin C loss.

MATH SCREEN

Vitamins are essential to life. They perform specific functions needed for growth and cell maintenance. Eating enough of the right foods to meet the United States recommended dietary allowances (USRDAs) for various vitamins are important to good health and well-being. The amounts recommended represent an estimate of the nutrients needed every day by healthy people.

The following table lists the vitamins found in fruits and vegetables and the corresponding USRDA for adults and children over four years of age.

Vitamin	USRDA
Vitamin A	5000 IU*
Vitamin D	400 IU
Vitamin E	30 IU
Vitamin C	60 mg
Vitamin B_6	2.0 mg
Thiamin (vitamin B_1)	1.5 mg
Folacin (folic acid)	0.4 mg
Biotin	0.3 mg

*An international unit (IU) is a form of measurement for fat-soluble vitamins A, D, and E. These vitamins occur in different biological forms. The IU is a standard measurement for all forms.

Use the following chart to answer the questions in this Math Screen.

Food	Approximate Vitamin A Content Per Ounce	Approximate Vitamin C Content Per Ounce
Baked sweet potato	2307 IU	6 mg
Cooked mustard greens	1624 IU	13 mg
Raw apricots	722 IU	3 mg
Orange juice	62.5 IU	15 mg

Study this sample problem to see how to translate USRDAs into amounts of food: How many ounces of orange juice are needed to meet the USRDA for vitamin C?

USRDA for vitamin C = 60 mg

One ounce of orange juice = 15 mg of vitamin C

$60 \div 15 = 4$ oz

Answer: 4 ounces of orange juice are needed daily to meet the USRDA for vitamin C.

Here is another sample problem for you to study: How many ounces of cooked mustard greens need to be consumed daily to meet the USRDA for vitamin A?

USRDA for vitamin A = 5000 IU

One ounce cooked mustard greens = 1624 IU of vitamin A

$5000 \div 1624 = 3.07$ oz

Answer: 3.07 ounces of cooked mustard greens need to be consumed daily to meet the USRDA for vitamin A.

Now test your math skills: How many ounces of raw apricots are needed to meet the USRDA for vitamin C?

SUMMARY

Vegetables and fruits are excellent additions to any meal. They add a variety of colors and textures that make food appetizing and appealing. In addition, they are relatively inexpensive; their cost is low in comparison to meat, fish, chicken, and dairy products. Vegetables and fruits are versatile as well. They can be served as entrées, accompaniments, appetizers, desserts, beverages, or garnishes. Some vegetables are used to season other foods.

Most importantly, vegetables and fruits are rich in nutrients—especially vitamins A and C—and low in calories. Included in a daily diet, they contribute to good health and help people on reducing diets to reach and maintain their ideal body weights.

A wide range of vegetables are available, including flower, fruit, green leafy, seed, root, and stem vegetables. Fruits come in three main varieties: summer, winter, and tropical.

To ensure quality, food service managers or buyers should only purchase those fresh and canned fruits and vegetables that have been graded at suitable levels. Proper storage, preparation, and cooking techniques help to maintain freshness, preserve nutrients, and avoid the color changes that can result from excessive heat or alkaline water.

QUESTIONS

1. What are the four natural pigments found in vegetables? What are the color characteristics of each? Give an example of a vegetable that contains each kind of pigment.
2. Acidic solutions have a pH less than what number?
3. Alkaline solutions have a pH greater than what number?
4. How should white or colorless vegetables be cooked to prevent color changes?
5. What color do green leafy vegetables turn if cooked in an acidic solution?
6. What are the six vegetable classifications? Provide examples of each.
7. Fruits are divided into what three classifications?
8. What are the four components of a salad? Give brief descriptions of each.
9. What is meant by *enzymatic browning?* What types of foods does it affect? How can it be prevented?
10. What two vitamins are found plentifully in most fruits and vegetables?
11. Why is the proper storage of fruits and vegetables so important? List six reasons.

ACTIVITIES

Activity One Ask your instructor to purchase the ingredients needed for the Greek salad recipe at the end of this chapter. Prepare the recipe, and evaluate its appearance, texture, and taste.

Activity Two Secure the ingredients needed for the molded fruit salad recipe at the end of this chapter. Prepare the salad, cooling it overnight to set. The next day, evaluate the salad's appearance, firmness, texture, and taste.

RECIPES

Greek Salad

Number of Servings: 24

Prepared Ingredients	Weight	Measure	Method
Washed and chopped romaine lettuce		3 heads	In a large bowl, combine all the ingredients and toss.
Tomatoes, cut in wedges		6 large	
Cucumbers, sliced		2	
Feta cheese, cut in cubes	½ lb		
Red onion, sliced in rings		1 medium	
Black olives, drained	16-oz can		
Dressing:			Add dressing and toss well until the salad is well coated.
Olive oil		½ cup	
Red wine vinegar		¼ cup	
Oregano		½ tsp	
Salt		¾ tsp	
Pepper		¼ tsp	

Serving Size: 1 cup.

Serving Suggestion: Add the dressing at the last possible moment and serve the salad immediately.

Molded Fruit Salad

Number of Servings: 16

Prepared Ingredients	Weight	Measure	Method
Unflavored gelatin		3 env	Dissolve gelatin in water and juice. Place over low heat and stir until gelatin is completely dissolved, making sure that all granules on the sides of the pan are scraped down into the liquid. Remove from heat. Add ice cubes to chill quickly. Add fruits and pour into mold. Stir occasionally during first two hours of chilling to prevent fruit from settling to bottom of mold. Refrigerate four hours.
Frozen fruit punch	12-oz can		
Water	20 oz		
Ice cubes	12		
Sliced strawberries		½ cup	
Seedless grapes		½ cup	
Sliced peaches		½ cup	To unmold, dip mold briefly in warm water to depth of gelatin. Loosen mold edges with knife. Place prepared serving dish on top of mold and turn upside down. Shake mold slightly to loosen gelatin. Garnish with greens.

Serving Size: 4 ounces.

Serving Suggestions: Serve on bed of lettuce alone or as an accompaniment.

17 GRAINS

VOCABULARY

bran
endosperm
germ
calorie
semolina
gluten
parboiled
enriched
potassium bromate
roux
whitewash
beurre manié
leavening agent
carbon dioxide (CO_2)
fermentation
proofing
rancidity
oxidized

OBJECTIVES

After studying this chapter you should be able to do the following:

- Identify the major grains grown for food and recognize the different components common to all grains
- Identify the products made from whole wheat grain
- Describe the different types of flour and the uses of each
- Recognize the types of leavening agents commonly used in baking
- Identify the major nutrients found in whole wheat or enriched grain products

Grains are a major component of the human diet, not only in the United States but throughout the world. In fact, they are the primary source of food in some countries. Food products made from grains are economical and nutritious, and they supply the body with a large amount of its needed energy. Grain products are an excellent source of complex carbohydrates.

The main types of grain are wheat, corn, rice, oats, barley, and rye. Many food products are made from these grains, including pasta, cereals, flour, and alcoholic beverages. With the exception of alcoholic beverages, whole grain and enriched products are important elements of a healthy meal, whether served as breads or as part of an entrée, soup, sauce, or dessert. The proper storage of grain products helps to maintain their quality and prevent the infestation of pests.

GRAINS AND GRAIN PRODUCTS

All grains have basically the same structural characteristics in their natural forms. These characteristics are illustrated in Figure 17–1.

The **bran** is the outermost layer and protective coating of a grain. It contains cellulose, some protein, thiamin, and minerals. This layer is the grain's best source of fiber. The softest component of a grain is the **endosperm.** It contains all of the grain's starch. The **germ** is the innermost layer. Rich in protein, fat, and B-complex vitamins, the germ is the part of the grain that reproduces; it therefore contains concentrated amounts of nutrients to support new growth.

Types of Grain

The most common varieties of grains include wheat, rice, corn, oats, barley, rye, and triticale. The following sections describe each type of grain and the many grain products produced from each.

Wheat

Wheat is the number one grain consumed throughout the world and in the United States. Wheat grain is sold as flour to be used as the main ingre-

FIGURE 17–1 Structural Characteristics of Grains

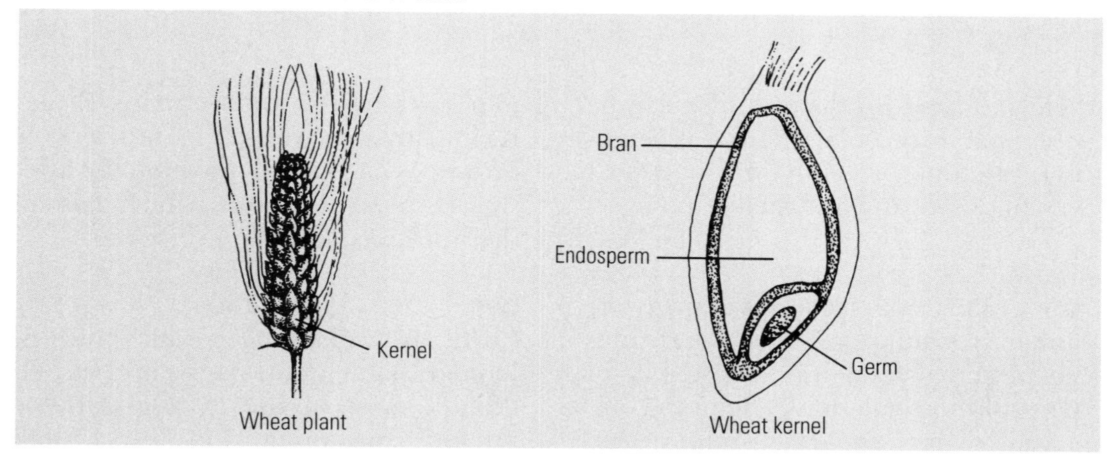

NUTRITION NOTE FOOD AS ENERGY FOR THE BODY

The human body needs energy to stay alive, to metabolize all body processes, to grow, and to perform physical activity.

To maintain an ideal body weight throughout life, an individual must achieve a balance between energy intake and energy output. Body energy is measured in calories. One **calorie** is the amount of heat required to raise one kilogram of water one degree Celsius. Body energy comes from foods containing carbohydrates, protein, and fat. Carbohydrates contain four calories per gram, as does protein. Fat contains nine calories per gram. If the calories consumed daily are more than the body needs, the excess is stored in the body as fat. Young men, for example, need approximately 1800 calories daily for metabolism, while young women need slightly less than 1600 calories. These amounts do not include the extra calories needed for physical activity.

Exercise is a major factor affecting body energy and caloric needs. The type of exercise and the length of time that one exercises determines how much food one needs to consume. Exercise should be done regularly in combination with eating nutritionally balanced meals. The following list shows different activities and the calories needed to perform them:

Type of Activity	Calories Needed (per Hour)
Underactivity level—sitting, reading, and eating	80–100
Light activity level—cooking and slow walking	100–150
Moderate activity level—sweeping, gardening, walking at a moderate pace	150–250
Vigorous activity level—walking fast, moderate aerobic dancing	250–350
Strenuous activity level—swimming, playing tennis, bicycling, playing soccer, and advanced aerobic dancing	350–500

dient in baked products, and it can be processed for cereals and for **semolina,** a high-protein durum wheat product used to make quality pasta.

Rice

Rice grain ranks second in worldwide human consumption and constitutes the principal food for almost half of the world's population. The bran component of rice contains protein, vitamin B complex, and vitamins E and K. Quality rice with all its nutrients is brown. White rice is nutritionally inferior because the bran, which has all the nutrients, has been removed.

Corn

Corn, also called maize, ranks third in United States grain consumption. Although most of the corn grown for human consumption in the United

Carbohydrates are by far the best-known fuel for the body. The human brain depends exclusively on carbohydrates as its energy source. Carbohydrates exist in two forms, simple and complex. Simple carbohydrates are those derived from the sugar in different foods such as table sugar, jams, jellies, syrup, honey, and fruit. Complex carbohydrates are those derived from the starch in foods like cereals, pasta, rice, potatoes, and flour products.

Protein is an energy source found in expensive foods, like meats. Its primary function is to build and repair body tissues. Fat rates inferior to carbohydrates as an energy source because it cannot be used efficiently by the brain and nervous system.

Carbohydrate foods have an unfair reputation for being fattening. What is actually fattening is the added butter on bread or the added sour cream on pasta and baked potatoes. For example, one slice of bread contains approximately 70 calories. When it is eaten with one tablespoon of butter spread on the top, the calories go up to 170. Foods from the grain, fruit, and vegetable groups contain complex carbohydrates that are rich in nutrients and fiber and should accompany every meal. Dietary fiber is the indigestible carbohydrate component of food that provides the necessary bulk in a diet and aids in elimination. Fiber-rich foods include fruits, vegetables, and whole-grain cereal products. People requiring low-calorie diets should reduce the consumption of fats and simple carbohydrates, such as those found in most sweet desserts.

States is served as a vegetable, the whole kernel can also be ground into corn meal and corn flour or processed for breakfast cereals. In addition, a specialized variety of corn kernels is used as popcorn, a popular snack.

Oat

Oat grain is mainly grown to feed cattle and horses. For human consumption, oat grain is made into breakfast cereals that need little or no cooking. It is also used as a stabilizer and a preservative.

Barley

Most barley grain is grown to feed animals. It is, however, also used for malting beverages like beer. Barley is not a popular grain for human consumption. In cooking, barley is mainly added to soups and cereals.

Rye

Rye grain is used to make bread products and crackers. It is also distilled into alcoholic beverages.

Triticale

Triticale is a newly discovered high-protein grain that is low in **gluten,** an elastic protein that provides strength and structure to dough and has a mild rye flavor. When combined with wheat flour, triticale makes tasteful, high-protein bread products. The grain is a cross between rye and wheat and is available in flour and flake forms. Figure 17–2 illustrates the different types of grains in their natural forms.

MARKET FORMS OF GRAIN

Grains are processed into a wide variety of foods, from breakfast cereals and pastas to rice and flour products. These grain products can be served as part of entrées, soups, sauces, and desserts, and they are valuable sources of B vitamins and iron.

Breakfast Cereals

Breakfast cereals are available in ready-to-eat or ready-to-cook forms. Cereals contain many important proteins; but cereal proteins, by themselves, are incomplete. The proteins in milk, however, complement cereal

ALCOHOL AWARENESS

Alcoholic beverages made from grain products include whiskey, vodka, beer, and gin. Whiskey, vodka, and gin are called spirits and are produced through processes called fermentation and distillation. Beer is produced by fermentation alone.

During fermentation, the sugar in grain is transformed into alcohol and carbon dioxide gas, which is activated by yeast. The carbon dioxide escapes, and the alcohol remains in the liquid.

When spirits are produced, the alcohol is separated from the fermented liquid. This process is called distillation.

The grains commonly used to produce alcoholic beverages are corn, barley, and rye. They are ground and then mixed with water and heated so that needed materials can be extracted. This process is called mashing. The criteria used to produce quality spirits and other alcoholic beverages are the selection of good grain products, the method of distillation, the length of the aging period, and the skill used in the blending procedures.

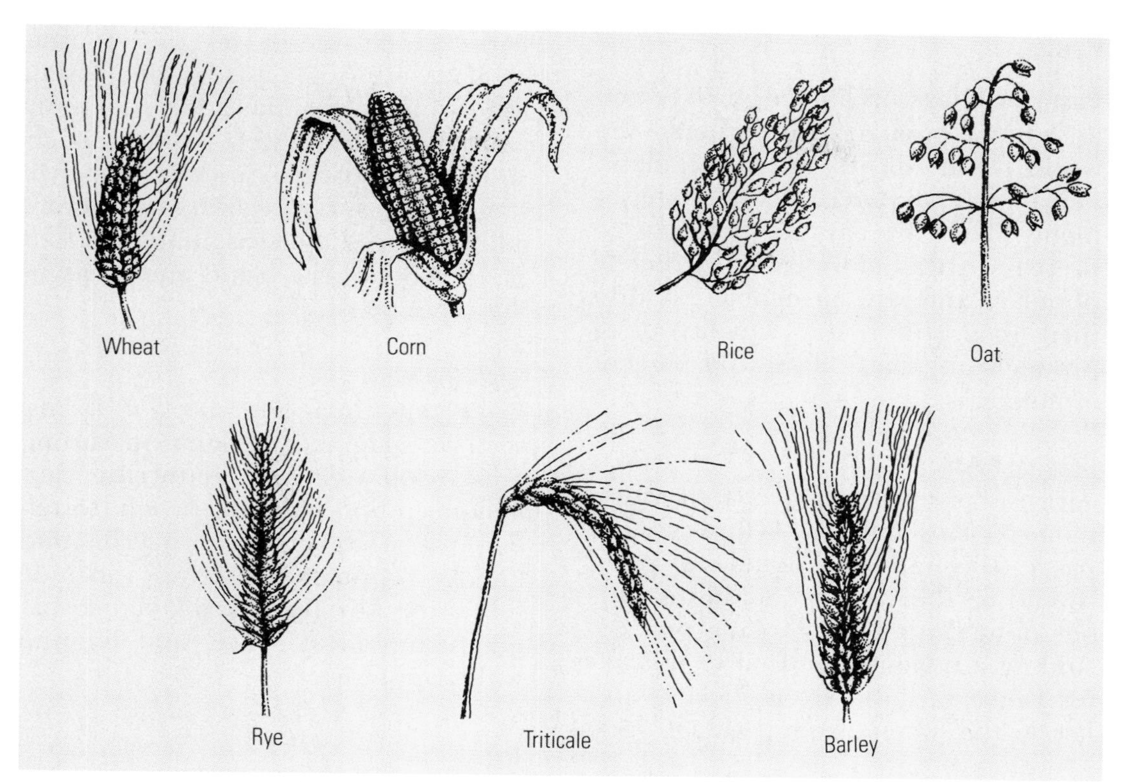

FIGURE 17–2 Commonly Used Grains in Their Natural Forms

proteins; when they are served together, all of the essential amino acids, or protein building blocks, are present. In addition, ready-to-eat breakfast cereals are fortified with 25 percent of the USRDA for seven vitamins and iron. Fortified products have had nutrients added to the vitamins and iron they contain in their natural states. Whole grain cereals contribute significant amounts of riboflavin, thiamin, niacin, phosphorus, and iron to the diet. One ounce of cereal is considered one serving.

ILLUSTRATION 17-1 Breakfast cereals are available in ready-to-eat or ready-to-cook forms.

Pasta

Pasta products originated in Italy and are now popular throughout the world. Pasta dishes are ideal menu items for commercial food establishments. They are easy to cook, can be prepared ahead of time, are extremely popular, and are profitable because their food cost is low. Quality pasta products are made from 100 percent semolina.

Cooking Pasta

All pasta products are prepared with a moist-cooking method. When cooking pasta, the ratio of liquid to pasta should be four to one. Salting is usually done during the cooking process. Adding a tablespoon of oil to the water prevents pasta from sticking and keeps the water from boiling over.

ILLUSTRATION 17-2 Pasta products originated in Italy and are now popular throughout the world.

Pasta should never be rinsed after cooking. Rinsing destroys the pasta's water-soluble vitamins and rinses away some of its flavor. Most pasta products should be cooked to a tender but firm stage and must be well drained before adding sauces and seasonings. Cooked pasta should taste bland and pleasant but never pasty.

Pasta Varieties

Pasta comes in a wide variety of forms, such as spaghetti, lasagna, macaroni, mostaccioli, manicotti, fettuccine, cannelloni, rotini, linguini, ravioli, tortellini, macaroni shells, vermicelli, and noodles. Popular pasta dishes include fettuccine Alfredo, baked lasagna, spaghetti and meat sauce, and linguini with clam sauce.

Rice

Rice grows in either long or short grains. It can be purchased in white (or refined), polished, converted, brown, or wild forms. In the past decade, rice dishes have gained popularity in the Western world. Rice is served as an accompaniment to main entrées as well as in soups, salads, desserts, and snacks.

White, or Refined, Rice

White, or refined, rice has had its bran and germ removed through a refining process. This process, however, also removes most of the nutrients. The only part that remains is the endosperm, which is made up of starch.

ILLUSTRATION 17-3 Rice can be purchased in wild (center, then clockwise from left), white, brown, converted, and polished forms.

Polished Rice

Polished rice is white rice with sugar and talc added to improve its luster. This rice must be well rinsed before cooking to remove all traces of talc. Polished rice cooks in 20 minutes using a moist method of cooking. This type of rice is popular with food service customers.

Converted Rice

Converted rice has a creamy white color. It has more nutrients than white rice but takes longer to cook. Converted rice cooks in 30 minutes by a moist method of cooking. The kernels retain their shape during the cooking process.

Brown Rice

Brown rice is often parboiled to speed the cooking process. A **parboiled** food item has been partially cooked in water. Parboiling rice permits nutrients from the bran and germ to enter the softer part of the grain, the endosperm. The rice is then allowed to dry, and the bran and germ are removed. Because it undergoes less processing, brown rice is much more nutritious than white or converted rice. It does, however, take longer to cook. Brown rice is rich in iron and B-complex vitamins, including thiamin, riboflavin, and niacin. Recognition of the nutritional value of brown rice has led to increased consumption.

Wild Rice

Wild rice is not a true rice but the seed of wild grasses. It is very expensive and considered a delicacy. In cooking, wild rice is almost always used in combination with other rice products. The nutritional value of wild rice is equal to whole wheat grain products.

Cooking Rice

Rice can be cooked using a variety of methods. All methods, however, must include the addition of a liquid. One cup of rice to two cups of liquid is the standard ratio for cooking rice. The volume of rice triples after cooking. One cup of uncooked rice yields three cups of cooked rice, which makes six one-half cup servings. When cooked, shorter grained rices tend to become more sticky than longer grained rices.

Flour

Flour is a fine, powdery substance that makes up the main structure of baked products. Flour is rich in starch and contains protein, fat, and minerals. Wheat flour is the most widely used flour in bread and dessert products. It is also the preferred flour for thickening soups, gravies, and sauces.

Technological innovations in the early twentieth century made machinery available to mill wheat kernels into white flour. Through the milling process, the bran and germ were removed, leaving only the endosperm available for flour. As a result, in the mid-1930s many Americans suffered vitamin deficiencies, in part from the loss of certain nutrients in flour. In 1941 the Enrichment Act required that flour be **enriched**—that the iron, niacin, thiamin, and riboflavin that were lost through the milling process be put back in the flour to enrich it to the previous level in the whole wheat. Federal law currently requires that white flour, pasta, cereal, and most bakery products shipped across state lines be enriched.

Types of Flour

Many types of flour are available. The following list describes the most commonly used varieties:

- Bread flour. Bread flour is a hard flour with a high gluten content. It is used to make bread. This flour has a creamy white color and granular texture, and it contains the highest amount of protein of all flours. It is milled from the hard varieties of wheat.
- All-purpose flour. All-purpose flour is a blend of soft and hard wheats. This flour is not for any particular product; it can be used to make many different baked goods, including hot breads.
- Cake flour. Cake flour is made from soft winter wheat and has the least amount of protein. Its tender, pliable gluten offers low resistance to the rapid expansion of cake batter when it rises. Cake flour is used to prepare cakes and other cake-based desserts.
- Bromated flour. Bromated flour is flour with potassium bromate added. **Potassium bromate** is a chemical used to make flour more stretchy and pliable for easy handling. Bromated flour is good for making bread products.

Flour as a Thickening Agent

Flour is a key ingredient in a number of thickening agents. The following mixtures are used in soups, sauces, and gravies:

- Roux. **Roux** is a mixture of flour and fat in equal proportions. The fat used may be butter, margarine, oil, chicken fat, or bacon fat. The flour can be from rice or wheat. Roux is added to soups and sauces to thicken them. It can be made ahead of time and kept in a refrigerator for convenience.
- Whitewash. **Whitewash,** sometimes called slurry, is a thin mixture used to thicken gravies and soups. It is made with cold water, flour, and

milk or stock, which are blended well to a thin, creamy consistency without visible lumps.

- Beurre Manié. **Beurre manié** is a mixture made with equal amounts of flour and butter. This mixture is not cooked; it is kneaded together and then stored in a refrigerator. Beurre manié is used to thicken thin sauces.

Other Ingredients Used with Flour in Baking

Although flour is the key ingredient used in baking, several other ingredients are also important. These additional ingredients include liquids, fats, sweeteners, salt, and leavening agents (air, steam, yeast, and chemical leaveners, for example). A **leavening agent** is a substance that helps baking products rise.

- Liquids. Liquids, such as water, milk, or fruit juice, are added to flour to moisten it and to ease the blending or mixing process. Adding water produces a baked item with a hard or crunchy texture, like French bread. Milk adds to a baked product's nutritional value as well as giving it a smooth texture and additional flavor. Fruit juices give flavor and help leavening agents to act.
- Fats. Fat softens a flour's gluten strands and thus makes a softer and smoother dough. Softer dough rises faster. Many different types of fat can be used in baking. Vegetable shortening is used because of its creamability and spreadability. Butter is creamable and flavorful. Other fats added to flour include margarine, lard, and oil.
- Sweeteners. Sugar is the sweetener of choice for baking products. It is available in ultrafine, very fine, and coarse forms, and in white and brown varieties. A recipe usually specifies what type of sugar to use, depending on the item being baked. Other sweeteners used in baking are syrup, molasses, and honey.
- Salt. Salt is used in baking for its taste. Salt also gives strength to the gluten structure. From the standpoints of texture, flavor, and health, an excess of salt should be avoided.
- Air. Air is added to baked goods as a leavening agent. It can be incorporated into dough or batter through the creaming or beating of shortening, the folding of egg whites, or by the sifting of flour.
- Steam. Steam is another leavening agent. It is created when water is heated. Baking a product that contains water inside an oven creates steam, causing the product to rise in volume.
- Yeast. Yeast, the most well-known leavening agent, is a live plant that grows with the addition of starch, sugar, and moisture under controlled temperatures. Yeast products rise best in temperatures between 84°F and 120°F. There are two types of yeast used in baking: dry and fresh. Dry yeast is dark

and has a granular texture. It is sold in small, sealed envelopes and dated with an expiration date to ensure quality. Fresh yeast, also called compressed or cake yeast, is sold in square-inch blocks. When quality yeast is added to baking products it produces **carbon dioxide (CO_2),** a colorless, odorless gas that makes dough rise.

- Chemical leaveners. The most common chemical leaveners are baking powder and baking soda. Both of these leaveners need heat and moisture to produce carbon dioxide. Each one, however, reacts uniquely with different ingredients.

POPULAR BAKED PRODUCTS

Grains are the key ingredient in most baked products. These products include breads, muffins, biscuits, cakes, pie pastries, and cookies.

Breads

Bread is an important accompaniment to a meal. The main ingredients for baking bread are flour, water, salt, and yeast (as a leavening agent). Popular types of bread include loaves, rolls, bagels, pitas, tortillas, and croissants. They can be made from white, whole wheat, or rye flours, or a mixture of them. Breads and other high-carbohydrate foods are low in calories. A slice of bread, for example, has approximately 70 calories.

Yeast breads are fermented until they double in volume. This process takes approximately two to three hours. **Fermentation** is the period when the yeast feeds on the starch and sugar, producing CO_2 gas. The final rising of the dough, after shaping but before baking, is called **proofing.** This process requires a warmer temperature than fermentation.

Muffins

Muffins are small cup-shaped breads that may be sweetened or unsweetened. They are usually served hot. Muffins often are made with the flavors of different fruits and nuts. Popular flavors are bran-nut and blueberry. When preparing muffins, the batter should be mixed with a

ILLUSTRATION 17-4 Grains are the key ingredient in most baked products.

MATH SCREEN

The weight of a person and the size of their muscles directly influences the number of calories they require daily. A young person with an inactive life-style needs far fewer calories to maintain her or his weight than a person engaged in strenuous activities, who needs additional calories to maintain weight.

One pound of body fat is equal to 3500 calories. Therefore, if a person wants to reduce his or her weight, that person needs to cut down his or her daily calorie consumption by 500 calories for a period of seven days in order to lose one pound of body weight.

500 calories × 7 days = 3500 calories

The following example shows how to calculate the number of days it would take to lose one pound of body weight by cutting down a person's daily caloric level by 500:

3500 calories ÷ 500 calories a day = 7 days

Test your math skills: How long would it take to lose ten pounds of body weight if a person reduces her or his calorie level by 250 calories per day?

spoon until the ingredients are blended. Overmixing causes tunnels inside the muffins and uneven peaks on their tops. Tunnels and peaks may also be caused by using a hard flour with a high gluten content or by using too many eggs. All-purpose flour is ideal for muffins because it contains mixtures of high- and low-gluten flours. A standard muffin should be moist, tender, golden brown, evenly grained, and free from tunnels, with a slightly rounded, bumpy top and a nut-like flavor.

Biscuits

Biscuits are shortened bread that is leavened with baking powder or baking soda. They are made from an unsweetened, stiff dough that has a lower ratio of liquid to flour than muffin batter. The pastry method is used to make biscuits. In this method, solid fat is cut into the dry ingredients, and liquid is added to form the dough. Hot, quality biscuits are flaky and should separate into layers when pulled apart. The top should be flat, the sides straight, and the crust golden brown.

Cakes

Cakes are made from mixtures of cake flour, liquid, eggs, sugar, chemical leaveners, and other ingredients. They are often made using a conventional method that involves creaming together the fat and the sugar and then adding the beaten eggs. In the final step, the sifted dry ingredients are added alternately with the liquid until all the ingredients are blended.

Pie Pastries

Pie pastry is made with all-purpose or pastry flour, salt, fat, and water. When baked, the pastry is slightly leavened by the steam produced from the small amount of water in the dough. Solid fats are best for making flaky pie pastry. Hydrogenated shortening is almost always the type used. The fat is cut into the flour and left in globules the size of peas. Pastry chefs prefer to use pastry flour because it is low in gluten and makes a tender crust. All the water is sprinkled at one time over the fat and flour mixture, which is then blended with a fork and gathered together gently with the fingers. Adding too much water and overmixing toughens the pie dough.

After the dough is mixed, it is kneaded lightly and left to set for about 30 minutes. This setting period allows the moisture to spread throughout the dough and the gluten to relax. Pie pastry should be rolled out on a lightly floured parchment paper or wax paper, which makes it easier to place the pastry in a pie pan.

Cookies

Cookies are different variations of shortened cake made with all-purpose flour, sugar, eggs, chemical leaveners, and flavorings. Cookies have a higher proportion of shortening and flour and a lower liquid and sugar content than cake. The liquid usually comes from the eggs. Cookie dough may be stiff, to make roll-type cookies, or soft, for drop, bar, or pan cookies.

STORING GRAIN PRODUCTS

Insects and rodents are the biggest problems in food storage areas. Every precaution must be taken to ensure that these pests do not contaminate the food supply. Pasta, rice, cereal, flour, and other dry-storage products must be stored in air-tight containers. Tightly sealed hard plastics, metals, and glass are the ideal materials to use. Paper products do not keep these insects and rodents out.

Grain products with a high fat content, like wheat germ, should be stored in a refrigerator to prevent them from becoming rancid. **Rancidity** is the development of offensive odors and flavors when the fat in a food is overexposed to oxygen, or **oxidized.** The proper storage of grain products eliminates problems with pests and rancidity, thus preventing unnecessary food costs.

SUMMARY

Grains are grown for human consumption and for feeding livestock. Today grain products include a variety of foods available in many different market forms. Pasta, rice, cereals, and flour are the market forms of grain from which thousands of menu items are prepared.

Foods prepared from grain are an essential part of a healthful daily diet because grains are a major source of energy for the human body. They also contain essential nutrients. During the mechanized milling process nutrients are lost, so flours are now enriched with iron, niacin, thiamin, and riboflavin.

There are several types of flour, including bread, all-purpose, cake, and bromated flour, used to prepare recipes. Flour is primarily used as a thickening agent and as a structural agent in products such as breads, muffins, biscuits, cakes, pastries, and cookies.

The proper storage of grain products is important to maintain their quality and prevent food loss from pest contamination. Grain products with a high fat content should be refrigerated to guard against rancidity.

QUESTIONS

1. All grains have what structural characteristics in common?
2. What are the main components of a grain's bran? endosperm? germ?
3. Which part of the wheat kernel has concentrated amounts of nutrients to support growth?
4. What is another name for high-protein durum wheat?
5. What seven grains are commonly used for food?
6. What is meant by the term *enrich?*
7. What is the difference between polished rice and converted rice?
8. What are the differences between bread flour and cake flour? Why are different flours used for breads and cakes?
9. What is the mixture of flour and water used to thicken gravies called?
10. What is the mixture of equal parts of butter and flour kneaded together called?
11. What five ingredients are commonly used in baking products in addition to flour? Explain their functions.
12. What four leavening agents are used in baking?

ACTIVITIES

Activity One Purchase the ingredients needed to prepare the recipes at the end of this chapter. Prepare the baked lasagna recipe. This activity may be accomplished in two class periods or two days by preparing the meat sauce, cheese sauce, and lasagna noodles the first day and then refrigerating them. Layer the lasagna and sauces the following day, bake, and evaluate the finished lasagna's color, texture, taste, and appearance. Remember that the hot meat sauce must be transferred into a shallow pan and refrigerated quickly to prevent the growth of bacteria.

Activity Two Prepare the rice pilaf with mushroom recipe. Make sure that the liquid and rice are measured accurately and are simmered no more than the time stated in the recipe. The finished product must be firm, and it should not stick together. Evaluate the finished product's color, texture, taste, and appearance with the help of your teacher.

RECIPES

Baked Lasagna

Number of Servings: 24

Prepared Ingredients	Weight	Measure
Ground beef	3 lbs	
Olive oil		½ cup
Onion, chopped		1 cup
Garlic, minced		3 cloves
Fennel seed		1 tsp
Oregano		1 tsp
Tomato paste	2 16-oz cans	
Water	1 qt	
Salt		1 tsp
Pepper		½ tsp
Cheese filling:		
Ricotta cheese	2 32-oz containers	
Parsley, chopped		½ cup
Whole eggs		6
Salt		¼ tsp
Pepper		⅛ tsp
Pre-preparation of noodles:		
Water		2 gal
Lasagna noodles	2 lbs	
Oil		2 tbsp
Salt		1 tbsp
Mozzarella cheese, grated	2 lbs	
Parmesan cheese, grated		1 cup

Method

In a large sauce pan, sauté chopped onion and ground meat in hot oil over medium heat until the meat is golden brown and is broken into small, pea-size pieces. Add garlic, oregano, fennel seed, salt, pepper, and tomato paste diluted in the water. Simmer on low heat for two hours, stirring occasionally. Sauce may need additional water during cooking.

Meanwhile: in a large bowl, combine ricotta cheese, parsley, and eggs. Mix well. Season with salt and pepper and refrigerate until meat sauce is ready.

Then, in a large stock pot, bring two gallons of water to a boil with salt and oil. Add lasagna noodles and cook until tender. Do not overcook. Drain.

Layering the lasagna: cover the bottom of 20″ × 12″ × 2″ baking pan with ⅓ of the lasagna noodles. Sprinkle with parmesan cheese and spread one third of the cheese filling over the noodles. Pour ⅓ of the meat sauce over the cheese filling and sprinkle again with parmesan cheese. Repeat until three layers of all mixtures have been added, ending with meat sauce on top. Sprinkle the top of meat with grated mozzarella and bake in a preheated 350° F oven for 45 minutes. Let stand out of the oven for 10 minutes before cutting. Cut lasagna in four sections vertically, or lengthwise, and six sections horizontally, or across, making 24 individual servings.

Serving Size: One 2″ × 3″ piece.

Serving Suggestion: Serve with a green tossed salad and garlic bread.

Rice Pilaf with Mushrooms

Number of Servings: 24

Prepared Ingredients	Weight	Measure	Method
Butter		¼ cup	Sauté mushrooms in butter over medium heat. Set aside. In large stock pot, bring stock to a boil with salt and pepper. Add rice and sautéed mushrooms.
Mushrooms, sliced	½ lb		
Chicken stock		1 gal	
Converted long grain Rice (white)		4 cups	Cover pot and simmer on low heat for 25 minutes.
Salt		2 tsp	
White pepper		½ tsp	

Serving Size: ½ cup.

Serving Suggestions: Serve as an accompaniment to meat, fish, or chicken dishes.

18
DAIRY PRODUCTS AND EGGS

VOCABULARY

pasteurization
homogenization
sweet acidophilus
Lactobacillus acidophilus
lactase
sweetened condensed milk
evaporated milk
emulsifying agent
osteoporosis
arteriosclerosis
hollandaise sauce
béchamel sauce

OBJECTIVES

After studying this chapter you should be able to do the following:

- Identify the different market forms of milk and their varieties
- Describe the structural components of an egg and the characteristics used to grade eggs
- Discuss procedures to follow for the safe storage of dairy products and eggs
- List and provide examples of the functions of eggs in food preparation
- Explain why dairy and egg products are important to the human diet

Milk, cheese, and eggs are valuable ingredients in any kitchen. In the past twenty years, cheese consumption has doubled. Milk and egg consumption has also increased. These foods are popular because they contain large amounts of essential nutrients, taste good, and are versatile.

Dairy products and eggs are the basic ingredients for hundreds of recipes, especially cream sauces, soups, custards, puddings, and desserts. They can be served at any time of the day and with any meal. Dairy and egg products are graded and stamped with a quality shield and dated to guarantee freshness. All these products must be stored at an appropriate refrigerator temperature to safeguard them from bacteria.

MILK

Milk is a liquid food that comes from female cattle, sheep, goats, camels, and other mammals. In the United States, cows' milk is the main type used by the dairy industry. Milk contains a large proportion of water plus many important nutrients. These nutrients include protein, fat, carbohydrates, riboflavin (and other B vitamins), calcium, and phosphorus. In addition, marketed milk is fortified with vitamins A and D. Vitamin D works with phosphorus and calcium to build and maintain healthy teeth and bones. Because of its nutrient content, milk is considered a valuable food for the growth and development of human life.

ILLUSTRATION 18-1 Because of their nutrient contents, milk and milk products are considered valuable foods for the growth and development of human life.

Milk Pasteurization and Homogenization

In the United States, milk sales are controlled by the federal government. For safety reasons, milk may not be sold in its natural form. All milk sold for public consumption must be pasteurized and homogenized. In **pasteurization,** the raw milk is heated at 165°F for 15 minutes. This process kills all the harmful bacteria that are naturally present in raw milk and prevents the transmission of illness-causing bacteria to human beings. In **homogenization,** the fat globules and fat-soluble vitamin A in raw milk are broken into many small particles and dispersed permanently throughout the product. This process results in a more uniformly textured milk.

Marketed Types of Milk

Fresh milk may be purchased in gallon, half-gallon, quart, pint, and half-pint plastic or heavy-paper containers. Restaurant establishments purchase fresh milk in five-gallon tube-fed containers, which are kept in special refrigerated dispenser units. Milk can also be purchased dried, canned condensed, and canned evaporated. The marketing and pricing of milk depends on the fat content. The higher the fat content, the higher the price of the milk.

Milk Forms

Milk is available in many forms. Use and storage are the two factors to consider when deciding which form to purchase.

Whole Milk

Whole milk is a pasteurized and homogenized milk with at least 3.25 percent milkfat content. It is used in sauces, casseroles, and desserts, as well as on cereals and as a beverage.

Two-percent Milk

Two-percent milk is a pasteurized and homogenized milk with 2 percent milkfat content. Two-percent milk is appropriate for low-fat, low-cholesterol recipes and low-fat beverages.

Skim Milk

Skim milk is a pasteurized milk with less than 0.5 percent milkfat. It is used in strict low-fat recipes and as a healthy beverage for those on low-fat diets.

Sweet Acidophilus Milk

Sweet acidophilus is a pasteurized milk with the bacterium ***Lactobacillus acidophilus*** added. This bacterium separates the milk sugar (lactose) into glucose and galactose for people who lack the enzyme lactase and thus cannot properly digest milk. **Lactase** is an enzyme found in the human intestine that helps metabolize the lactose in milk.

Sweetened Condensed Milk

Sweetened condensed milk is a sterilized milk with 40 percent added sugar and half its water removed. It is used mainly to prepare baking products, desserts, and dessert toppings. Sweetened condensed milk makes excellent candies because this type of milk caramelizes smoothly.

Evaporated Milk

Evaporated milk is a sterilized milk with half its water removed. It can be used full strength in baking products or as a cream on foods. Evaporated milk is canned and may be stored a long time without being refrigerated. Water can be added for drinking or cooking.

Dried Milk

Dried milk is a dehydrated milk. It is made from whole or skim milk that has been made into a powder form. The milk can be rehydrated with water for beverages or for cooking and baking products. If used as a beverage, dried milk tastes better after standing in a refrigerator several hours after rehydration.

Yogurt

Yogurt is made from milk. A special bacterium sours the milk and creates a soft, custard-like texture. Yogurt may be plain or flavored, and it can be eaten as a snack, dessert, or fruit dressing.

Cream Forms

Cream is the rich, high-fat part of milk. To be labeled cream, a product must meet certain minimum federal standards. Several types of cream are available in stores.

Whipping Cream

Whipping cream contains more than 30 percent milkfat. It is used to prepare pastries or the gourmet sauces that garnish many different desserts.

Light Cream

Light cream contains 18 to 30 percent milkfat. Sometimes called coffee cream, it is used to make white sauces, custards, puddings, and many other desserts.

Half-and-Half

Half-and-half, the lightest cream, is a mixture of whole milk and light cream. It contains 10.5 to 18 percent milkfat and is most commonly used to flavor hot coffee and hot tea. Half-and-half can also be used to prepare a variety of white sauces and creamy desserts.

Ice Cream

Ice cream is made with milk, light cream, sugar, flavorings, and, usually, stabilizers. Ice cream must contain at least 10 percent milkfat to meet federal regulations. Ice cream is judged on its flavor, appearance, texture, and melting quality. The appearance must be smooth and shiny, and the texture must be creamy and finely grained. The color depends on the type of flavor used. Ice cream should completely melt at room temperature. If ice cream holds its shape at room temperature, it is not a quality product.

Ice Milk

Ice milk has the same quality standards as ice cream and is required by the federal government to contain between 2 and 7 percent milkfat. Ice milk has a more grainy texture than ice cream because the milkfat content is lower; a higher proportion of sugar is required for a smooth product.

Butter

Butter is made from the fat derived from dairy cream, but it is not considered a true dairy product because it lacks the nutrients found in milk and cheese. Butter is an excellent source of vitamin A; it is, however, high in cholesterol. Butter is graded by the United States Department of Agriculture through the use of a scoring system that includes the following grades: U.S. Grade AA, U.S. Grade A, and U.S. Grade B. Butter ranked U.S. Grade AA is made from sweet cream. U.S. Grade A butter is made from cream that has been soured with added neutralizers. U.S. Grade B butter is also made from milk that has soured, but it has more neutralizers and contains more salt than U.S. Grade A butter.

Butter is sold in salted or unsalted forms. Salt adds flavor and maintains the quality of butter longer. Vegetable dyes are added to give butter its uniformly yellow color. Butter contains 80 percent milkfat by weight. It is used as a spread, to season foods, and in baking and frying. When frying, care must be taken to prevent browning. Frying must be done on low heat. High and medium heat burns butter quickly and causes it to turn brown, which also changes its flavor.

CHEESE

Cheese is an increasingly popular food item that has gained wide acceptance because there are so many different varieties from which to choose. These varieties satisfy almost everyone's taste buds.

Today's finely honed cheese-making techniques were passed on from cheese makers through many generations. Immigrants such as the Germans, Italians, Greeks, Dutch, and Swiss brought cheese recipes to the United States. These immigrants produced the same cheese varieties that their ancestors made, and cheese soon became a popular food selection. Varieties of cheese include Swiss, Muenster, feta, Roquefort, ricotta, and Gorgonzola.

ILLUSTRATION 18-2 Cheese is an increasingly popular food item that has gained wide acceptance because there are so many different varieties from which to choose.

Cheeses may differ in texture and flavor, but cheese-making methods are similar among the different varieties. All cheeses are graded on the basis of flavor, body, texture, color, and appearance. Moisture and fat contents are state regulated. Important USDA inspections for cheese take place at the production level. The large shipping cartons that stores receive are also inspected and stamped with a grade.

Cheese belongs to the dairy group, and it is rich in many essential nutrients. Five quarts of milk are needed to make one pound of cheese. One ounce of cheese has 75 percent of the protein, calcium, and vitamin A found in one eight-ounce glass of milk.

Cheese is used to create attractive cheese trays and to flavor many delicious dishes. Popular ways to use cheese in meals prepared throughout the day are as follows:

- Breakfast—omelets, soufflés, and quiches
- Lunch—sandwiches and quiches
- Appetizers—cheese and fruit trays, cheese balls with crackers, breads, or cheese and wine combinations
- Dinner—main entrées such as baked lasagna, soups, and sauces
- Dessert—Danish, cheesecakes, and cheese and fruit trays

EGGS

Eggs supply a variety of nutrients and serve numerous functions in food

MATH SCREEN

Study the following chart, which shows the calcium contents in various dairy products:

Product	Portion Size	Calcium (mg)
Whole milk	1 cup	228
Skim milk	1 cup	296
Ice cream	1 cup	194
Cheddar cheese	1 ounce	213
Cottage cheese	1 cup	230
Yogurt, from whole milk	1 cup	272

The list that follows shows the recommended dietary allowance (RDA) for calcium in the diet of people in different age groups and situations:

- Adults—800 milligrams per day
- Teenagers—1200 milligrams per day
- Pregnant or lactating women—1200 milligrams per day

Examine this sample problem to see how to calculate how much of certain foods a person must eat to meet the RDA for calcium: How many cups of whole milk would a growing teenager need to consume daily to meet the RDA for calcium?

The total milligrams of calcium needed daily divided by the number of milligrams of calcium per cup of whole milk equals the number of cups needed daily.

1200 mg ÷ 228 mg = 5.2 cups of whole milk

Now test your math skills: How many ounces of cheddar cheese would an adult person need to consume daily to meet the RDA for calcium?

ILLUSTRATION 18-3 Cheese and fruit trays are popular appetizers.

preparation. When compared to other foods, eggs are inexpensive and thus cost-effective for food service establishments. Figure 18–1 illustrates the basic structural components of an egg.

The grading of eggs is done by the United States Department of Agriculture (USDA). Quality grades include USDA Grade AA (top grade), USDA Grade A (intermediate grade), and USDA Grade B (low grade). Grade AA eggs have clean shells that are free from cracks, air cells that do not exceed 1/8 inch in depth, and regular shapes. When cracked open, they have thick egg whites and yolks that stand high. The yolks should stand in semicircles, and the whites should be thick with two distinct layers of thickness. Grade A eggs are flatter and a bit more spread out than Grade AA eggs. These eggs may be used for frying or poaching. When Grade B eggs are broken, they spread out thinly and their yolks may break. These eggs are used for cooking and baking. Figure 18–2 shows the interior qualities of differently graded eggs.

The color of an egg's shell or yolk is not an indicator of quality. A dark yellow yolk is no more nutritious than a light yellow yolk. The interior color

FIGURE 18–1 The Basic Structural Components of an Egg

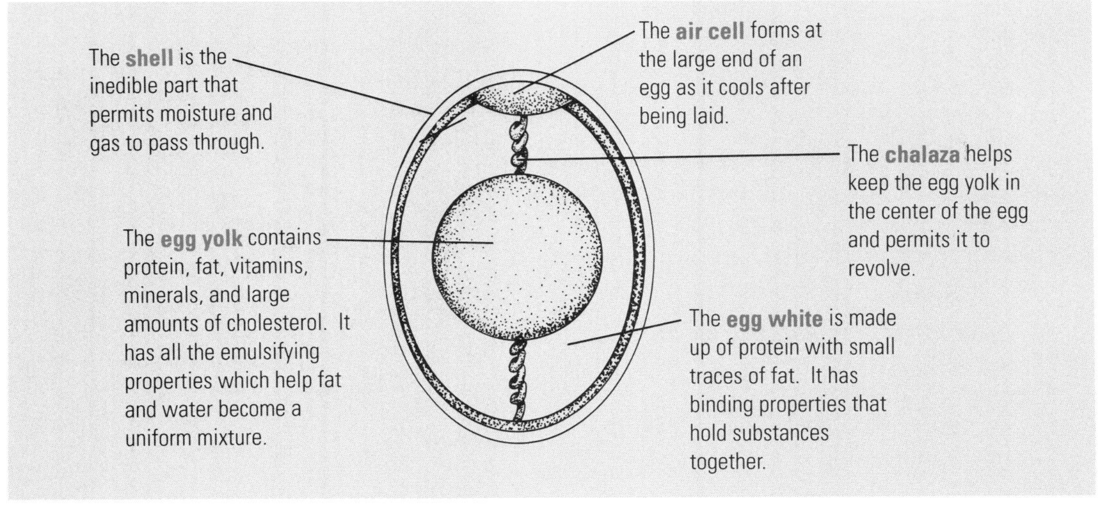

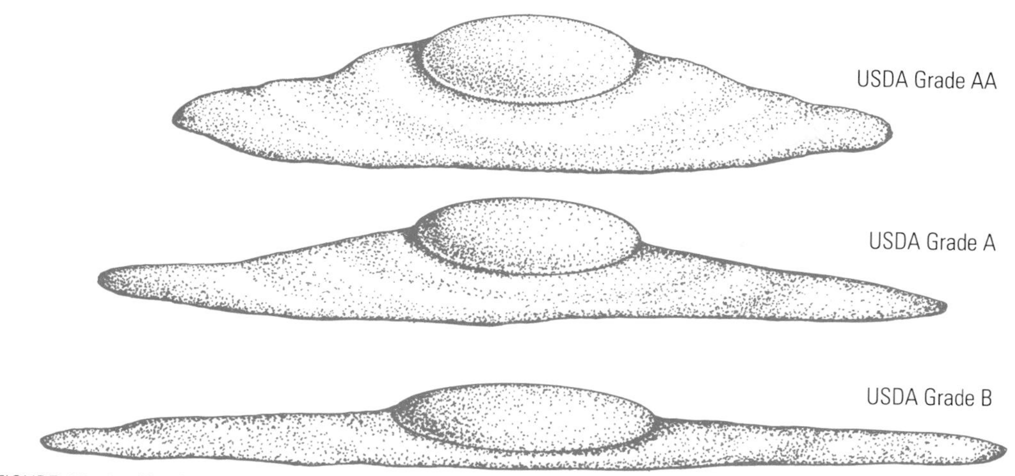

FIGURE 18–2 The Interior Quality of Differently Graded Eggs

of an egg is determined by the type of feed that the fowl consumed. The color of an egg shell depends upon the breed of hen that laid the egg. Eggs left at room temperature will lose their quality over a period of time. The air cell becomes larger, the yolk becomes watery and flat, and the egg itself becomes less flavorful. USDA Grade AA eggs, for example, can deteriorate in quality and become USDA Grade B if they are not stored properly.

Marketed Sizes of Eggs

Fresh eggs are marketed according to size and weight. They are usually sold by the dozen. Each dozen must be packed with 12 uniformly sized eggs. The carton is then weighed and priced accordingly, labeled for size, and stamped with a purchase date. The eggs' quality is guaranteed up to the date indicated on the stamp. The carton is also stamped with a grade shield (see Figure 18–3 on page 248) certifying that the eggs have been graded for quality and size.

Table 18–1 (page 248) lists the different sizes and weights given to eggs

ILLUSTRATION 18-4 Eggs supply a variety of nutrients and serve numerous functions in food preparation.

FIGURE 18–3 Grade Shield for USDA Grade A Eggs

by the dozen. Most recipes use large eggs. Therefore, if smaller or larger eggs are used, it is best to use weight instead of size to more accurately fulfill the recipe requirements.

Eggs and Their Uses in Food Preparation

Eggs are a popular breakfast item. In the United States, many people consider breakfast the most important meal of the day. Breakfast often consists partly of eggs that are poached, fried, scrambled, or in omelet form.

It is important to cook eggs at low temperatures and just until done. Eggs are high in protein, and proteins are heat sensitive. Proteins become tough when cooked at high temperatures. Before cooking an egg in its shell, the large end of the shell should be pierced with a clean pin to prevent the shell from cracking when it contacts hot water. Eggs should be simmered instead of boiled. Boiling causes eggs to become rubbery and to develop a dark line between the yolk and the white.

Like dairy products, eggs are used as the main ingredient in custards, puddings, and a variety of other baked products. Eggs are a popular item in food preparation because they serve several important functions. Eggs, for example, are often used as an **emulsifying agent**—an ingredient that helps combine two ingredients that do not mix well, such as oil and vinegar. These functions and examples of their use are listed in Table 18–2.

STORING DAIRY PRODUCTS AND EGGS

Milk, cheese, and eggs are favorable to bacterial growth because they contain protein and moisture, two ingredients that help bacteria to thrive. Milk containers and dispensers must be thor-

TABLE 18–1 Egg Sizes and Their Weights per Dozen

EGG SIZE	EGG WEIGHT PER DOZEN	NUMBER OF EGGS PER CUP
Jumbo	30 ounces	3
Extra large	27 ounces	4
Large	24 ounces	5
Medium	21 ounces	6
Small	18 ounces	7
Peewee	15 ounces	8

TABLE 18–2 Common Egg Functions and Uses

EGG FUNCTIONS	EXAMPLE OF USES
As a gelling agent	Custards
As a thickener	Sauces and puddings
As a structural component	Baking goods
As a leavening agent	Meringues
As a source of moisture	Cookie doughs
As an emulsifying agent	Mayonnaise
As a binding agent	Breadings or coatings (as in meat loaf)

oughly cleaned in hot, soapy water and rinsed with a sanitizing agent. Fresh dairy products can be stored at refrigerator temperatures from 32°F to 40°F until their expiration date. This date is usually stamped on the carton or plastic container. Dried and canned milk products have a longer storage life but must be stored in a dry, ventilated storeroom at temperatures from 50°F to 65°F. Nonfat dried milk has a longer storage life than dried whole milk. The fat content in dried whole milk speeds up the spoilage process by being susceptible to rancidity.

When storing cheese, it should be wrapped tightly or stored in an airtight container to preserve its moisture and flavor. Proper storage is especially important for strongly flavored cheeses such as Limburger, Brie, and blue. Fresh soft cheeses do not improve with age, but instead spoil quickly and need to be used as soon as possible. If mold develops on parts of a ripened cheese, the moldy parts can be cut off with a knife and the remainder of the cheese is still edible. Cheese must be served at room temperature for optimum taste and flavor, and thus should be taken out of the refrigerator an hour before serving. Cheese freezes well without compromising its nutritional value or flavor. The texture of frozen cheese changes when thawed, however, becoming more crumbly and difficult to slice into uniform pieces.

Cracked eggs are prime sources of bacterial growth and contamination. In addition to protein and moisture, cracked eggs give bacteria the oxygen they need to grow. Bacterial growth is especially dangerous when eggs are left out at room temperature.

Eggs should never be washed before storage. Their porous shells absorb the moisture and any bacteria that are present on the outside of the eggs. Raw and cooked eggs should be refrigerated in paper cartons to prevent them from drying out. If stored properly, raw eggs can keep their quality until the date that is stamped on their carton.

All parts of an egg may be frozen. Whole eggs must be beaten with some salt or sugar into a homogenous mixture before freezing. This process prevents the eggs from solidifying during freezing. Salt and sugar are added to lower the freezing point. Egg whites may be frozen in their natural state.

THE NUTRITIONAL VALUE OF DAIRY PRODUCTS AND EGGS

Dairy products are good sources of protein, calcium, phosphorus, riboflavin, vitamin A, and vitamin D. Because most foods do not contain vitamin D, fresh whole milk is fortified with this vitamin to prevent deficiencies. Vitamin D is important in regulating the minerals, such as calcium and phosphorus, that help form bones and teeth. Vitamin D deficiency causes rickets, a disease in which the bones become weak and deformed. Nonfat

NUTRITION NOTE OSTEOPOROSIS AND ARTERIOSCLEROSIS

Osteoporosis is a disease characterized by a degeneration of bone tissue. The bones become thin, porous, weak, and are easily fractured. Osteoporosis is more common in women than in men and strikes approximately 30 percent of people over the age of 65.

The major symptom of osteoporosis is bone loss over a period of time. Bone is made up of calcium, phosphorus, salts, and other minerals. Calcium and phosphorus, however, seem to be the major nutrients contributing to healthy and strong bones. Therefore, a deficit of calcium and phosphorus appears to be the major cause of osteoporosis. It takes years to develop this disease; a deficit of these minerals in the diet at a young age may not become apparent for many years.

Ninety-nine percent of the calcium in the body is found in bone and teeth. The other 1 percent is in the body fluids that nourish cells. Calcium regulates the transport of minerals, such as sodium and potassium, that are important to nerve transmission. It is also required for muscle contraction (especially of the heart muscle), helps blood to clot, and maintains the collagen that holds cells together.

The recommended dietary allowance (RDA) for calcium is 800 milligrams per day for adults and 1200 milligrams for children with growing needs. The main source of calcium is milk. Other dairy products, such as yogurt and cheese, are also good sources. One cup of whole milk contains 228 milligrams of calcium, one cup of yogurt contains 272 milligrams, and one ounce of cheddar cheese contains 204 milligrams.

Arteriosclerosis is a cardiovascular disease that attacks the arterial walls of the human body. It is the major cause of heart disease and heart attack in the United States. Cholesterol contributes to arteriosclerosis. The average American diet contains between 500 to 1000 milligrams of cholesterol daily. In countries where the average cholesterol intake is low (under 300 milligrams), heart disease is less prevalent.

All body cells have the capability of synthesizing cholesterol, but the liver is the organ primarily responsible for cho-

milk is fortified with vitamin A because, when the fat is removed from the milk, this fat-soluble vitamin is removed with it.

Eggs are an excellent source of protein, vitamin A, B vitamins, and iron. They also supply zinc, iodine, phosphorus, and magnesium. Eggs are especially valuable to vegetarians, helping them meet their protein and iron needs. All of these nutrients are important for the growth and development of the human body.

Eggs also contain large amounts of cholesterol, which can cause arteriosclerosis in certain individuals. **Arteriosclerosis** is a disease characterized by fatty deposits, or plaque, in the arterial walls. These fatty deposits can block the flow of blood, causing heart

lesterol output. This output depends on the amount of cholesterol ingested in the diet. If a healthy person does not consume the cholesterol she or he needs, the body has the ability to make the necessary amount. Some individuals have several hereditary abnormalities that keep their blood cholesterol levels high, regardless of diet. Most people with a cholesterol problem, however, can lower their blood cholesterol level by reducing the amount of cholesterol in their diet.

The major factors that contribute to arteriosclerosis are heredity, life-style, cigarette smoking, obesity, and diet. Studies show that, besides dietary cholesterol, blood cholesterol levels may be negatively affected by the excess consumption of saturated fats and sugar (in concentrated sweets), and by a lack of vitamins, minerals, and fiber.

Diet is only one of the causes of arteriosclerosis, but it is one that can be changed or modified. Although the average healthy person does not necessarily need a strict low-fat, low-cholesterol diet, it is wise to follow the Dietary Goals of the United States, proposed in 1977 by the U.S. Senate Select Committee on Nutrition and Human Needs, which include the following recommendations:

- Increase consumption of fruits, vegetables, and whole grains.
- Decrease consumption of foods containing large amounts of cholesterol.
- Decrease consumption of foods high in saturated fat, and substitute polyunsaturated and monounsaturated oils for oils made with saturated fats.
- Substitute low-fat and nonfat milk for whole milk (except for children).
- Decrease consumption of animal fats by not eating as much red meat and trimming all visible fat off the meat before cooking.
- Decrease consumption of refined sugars (in concentrated sweets).
- Decrease consumption of salt and foods containing large amounts of salt.

In addition, increasing one's activity level, reducing or eliminating cigarette smoking, and drinking alcohol in moderation all help to maintain a healthful life-style.

attacks or strokes. Because eggs contain high levels of cholesterol, the American Heart Association recommends eating no more than three eggs per week. It also recommends reducing the consumption of foods containing cholesterol and saturated fats, such as butter, lard, and other animal fats. These fats should be consumed in moderation as a preventive measure.

POPULAR FOODS PREPARED WITH DAIRY AND EGG PRODUCTS

A number of food items are made with a white sauce as their base. A standard white sauce is made by thickening one cup of milk with one tablespoon of flour. Cream soups and a variety of sauces and puddings can be made from a white sauce.

Hollandaise sauce is a basic egg sauce from which many other sauces are made, including maximillian, cherbury, mussolini, and bernaise sauces. Hollandaise sauce is commonly served over vegetables. Its primary ingredients are egg yolks and butter, with lemon juice or vinegar added.

Béchamel sauce is a basic milk sauce from which other sauces are made. Béchamel sauce is made with roux and hot milk. Sauces made from a béchamel base are Newburg, à la king, cheese, cardinal, and Mornay.

Pudding is made with milk and is thickened with cornstarch, eggs, or both. Pudding can be flavored with many different ingredients, such as banana, butterscotch, pineapple, chocolate, and vanilla.

SUMMARY

Dairy and egg products are important foods because they contain concentrated amounts of many nutrients needed by humans for good health and growth. Two nutrients in milk, for example, are not readily obtained from other sources: calcium and riboflavin. Cheese contains concentrated amounts of protein, calcium, and vitamin A, and egg products supply protein and iron.

Dairy products and eggs are marketed in a wide range of forms. Milk products are categorized according to their fat contents. Different cheese-making techniques, based on centuries-old recipes, produce the variety of cheeses available today. Eggs are sold by size and weight, and they are inspected and assigned grades by the USDA.

Dairy and egg products are necessary ingredients in many entrées. These entrées include sauces, soups, casseroles, desserts, and appetizers.

Sanitary refrigerator storage for dairy products and eggs is important to prevent contamination by bacteria. These foods are rich in protein and moisture, two factors needed for bacterial growth. By law, milk must be pasteurized and homogenized to kill harmful bacteria and produce a uniformly textured product. These products should be used before the expiration dates that are stamped on their cartons.

QUESTIONS

1. Whole milk contains what percentage of milkfat?
2. Ice cream contains light cream, sugar, flavorings, stabilizers, and what other ingredient?
3. What type of milk has *Lactobacillus* bacteria added to change the lactose structure for people with lactase deficiency?
4. What four major nutrients are found in dairy products? What two major nutrients are found in eggs?
5. What are the criteria used to grade all cheeses?
6. What are two factors in cheese making that are state regulated?
7. What are the five structural components of an egg?
8. What are the characteristics used to grade eggs?
9. What three functions do eggs fulfill in food preparation?
10. How should cheese be prepared for storage?

ACTIVITIES

Activity One Purchase the ingredients necessary for the preparation of the béchamel sauce and cheesecake and recipes at the end of the chapter. Prepare the béchamel sauce. Evaluate a small sample of the sauce, while it is hot, for color, texture, consistency, and taste. The sauce should be free from lumps, with a smooth and creamy consistency.

Activity Two Prepare the cheesecake. Cool overnight and then sample and evaluate the recipe. The characteristics for evaluation should include color, texture, consistency, taste, and appearance.

RECIPES

Béchamel Sauce (medium thickness)

Number of Servings: 24

Prepared Ingredients	Weight	Measure	Method
Butter	8 oz	1 cup	In a heavy saucepan, melt butter. Add flour, making a roux, and cook for five minutes.
Flour		½ cup	Add hot milk, stirring very fast with a wire whip until the mixture thickens and starts to
Milk, scalded		2 qts	bubble. Remove from heat and season with salt.
Salt		½ tsp	To remove lumps, strain through a china cap into a clean pan.

Serving Size: 3 ounces.

Serving Suggestions: Excellent over cooked vegetables or pasta dishes.

Cheesecake

Number of Servings: 24

Prepared Ingredients	Weight	Measure	Method
Pie shell:			
Graham cracker crumbs		3½ cups	In a large bowl, combine graham cracker crumbs, butter, walnuts, and sugar. Press mixture in equal amounts into two nine-inch springform pans; set aside.
Butter, softened		½ cup	
Walnuts, finely chopped		½ cup	
Sugar		¼ cup	
Filling:			In a large bowl, beat cream cheese, eggs, sour cream, vanilla, and lemon peel until thick and creamy. Pour equal amounts into the prepared pie shell pans. Bake in preheated 350°F oven for one hour or until the center of the cake is set and an inserted knife comes out clean. Place on wire rack to cool.
Cream cheese		3 lbs	
Eggs		6	
Sour cream		2 cups	
Lemon peel, grated		1 tbsp	
Vanilla		1 tbsp	
Strawberry glaze:			Place sliced strawberries in a saucepan with the water and cook for two minutes. Combine cornstarch and sugar and stir into the hot mixture. Cook until it starts to boil, stirring constantly, until mixture becomes thick and clear. Add food coloring, if needed, to brighten the color. Arrange whole strawberries equally on top of the two cheesecakes and pour the strawberry glaze topping on top of the whole strawberries. Cut each pie into 12 equally sized wedges.
Fresh strawberries, sliced		4 cups	
Fresh strawberries, whole		4 cups	
Water		1½ cups	
Cornstarch		2½ tbsp	
Sugar		1½ cups	
Red food coloring (optional)		2–3 drops	

Serving Size: One wedge of a nine-inch cheesecake.

Serving Suggestions: Serve with coffee or tea after a light meal.

PART SEVEN

MANAGEMENT CONTROLS

19
ENTREPRENEURSHIP

VOCABULARY

entrepreneur
creative imitation
innovation
market research survey
target market
customer profile
competition
trend
perceived value
American nouvelle cuisine

OBJECTIVES

After studying this chapter you should be able to do the following:

- Explain the term *entrepreneur*
- Discuss the three entrepreneurial keys to success
- Relate the three major theories of entrepreneurship to business management and marketing functions
- Outline the process for identifying and creating a need for a new product or service
- Discuss dining trends that are creating entrepreneurial opportunities in the food service industry

Entrepreneurship is the foundation of the American food service industry. Almost every food service business that exists in this country began as an idea that was developed by an entrepreneur. An **entrepreneur** is a person who manages and assumes the risks of a business. Through the application of basic business management principles, an entrepreneur organizes, manages, and assumes the risk of developing an idea into a business. Successful entrepreneurs have the ability to apply the theories of entrepreneurship when they recognize the need for a product or service and then develop, produce, and market it. An entrepreneur must also have the physical stamina and energy needed to develop and maintain any successful business enterprise.

Food service businesses have been developed to meet the need for different types of public eating facilities. As discussed in Chapter 1, every community since early times has recognized the need for prepared foods outside of the home. Throughout history, public food service facilities have become more and more sophisticated, leading to the variety of restaurants that exists in every major city today. The industries that have developed to supply food and equipment to these restaurants are equally varied and sophisticated.

Present-day entrepreneurs continue to identify products and services that are needed in the marketplace. This chapter covers some of the ways that these needs are identified, and how entrepreneurs successfully develop products and services to meet these needs.

ENTREPRENEURSHIP THEORIES

Just having a good idea is not enough to become an entrepreneur. Successful entrepreneurs make definite decisions about how to develop their ideas into saleable products and services. They lay out plans that apply one or more of the three basic theories concerned with business management and marketing functions. These three theories are leadership, creative imitation, and innovation.[1]

Leadership

An entrepreneur who uses the leadership theory is developing a product or service that has never before existed. The goal of the entrepreneur is to introduce the product or service and become the leader of that industry. To be successful, an entrepreneur applying the leadership theory must know who the customers will be and how to sell the service or product to them. The entrepreneur must know how to obtain financial backing, and she or he must have the business knowledge to produce the product or service and to operate the business.

Creative Imitation

The **creative imitation** theory takes a successful product or service and does something different to it, creating a new consumer market. The customers are already familiar with the product or service and are usually ready to accept another variation of it.

Innovation

An entrepreneur who recognizes a need and develops a product or ser-

[1] Peter F. Drucker, *Innovation and Entrepreneurship* (New York: Harper & Row, Publishers, 1985), 210, 220, 243.

vice to fill that need is using the theory of innovation. An entrepreneur applying the **innovation** theory creates potential customers by identifying a need, determining what product or service the customers want, and establishing how much they are willing to pay for the product or service.

The two theories that are most commonly applied to entrepreneurial food service ventures are creative imitation and innovation. One of the best examples of the application of these theories is McDonald's.

Ray Kroc, the founder of McDonald's, was a food service equipment sales representative in California. One of his customers owned a small neighborhood hamburger stand that sold a tremendous number of milk shakes in proportion to the size of its operation. Kroc identified the success of the milk shake sales with the owner's ability to produce consistently a quality product at a price that customers wanted to pay. He purchased the hamburger stand and went on to apply this technique to every food product that he sold, thus creatively imitating and innovating upon the original hamburger stand. By applying business management and marketing principles to his restaurant operation, Kroc developed a chain of hamburger stands that today is represented in countries around the world.

ILLUSTRATION 19-1 Ray Kroc, the founder of McDonald's, identified success with the ability to produce consistently a quality product at a price that customers wanted to pay.

McDonald's spends millions of dollars annually to conduct market research. These marketing activities include consumer surveys, working with consumer groups, and analyzing the new products that competitors are offering. New McDonald's product developments in the past few years have included Chicken McNuggets, dessert items, salads, and an expanded variety of hamburger combinations. Salads have been the most successful of these new product lines. Identifying their customers' desire to eat lighter, low-calorie foods in a fast-food setting, McDonald's concentrated its efforts to produce a line of salad items that would satisfy a wide variety of customer needs. The result was an initial offering of three salad bases: chicken oriental, chef's, and garden. The response was overwhelming in terms of both customer satisfaction and financial gain.

KEYS TO ENTREPRENEURIAL SUCCESS

In addition to applying entrepreneurial theories, McDonald's efforts illustrate the application of three keys to entrepreneurial success. These keys are identifying the target market, understanding the customer profile, and knowing the competition. Data are gathered from conducting market research surveys related to these keys. A **market research survey** is an examination of a specific group of customers, competitors, and target markets within an established geographical area. Successful entrepreneurs know who their customers are and understand what their customers want.

The Target Market

A **target market** is the group of customers who would be most likely to buy a product or service over a period of time. Unless entrepreneurs can identify their target markets, they will not be able to successfully determine different needs for new products or services.

The Customer Profile

A **customer profile** consists of a series of general characteristics that identify the target market to which a customer belongs. Age, occupation, geographical location, average annual income, family size, level of education, and spending patterns are the major characteristics that make up a customer profile. In food service research, questions related to what type of restaurant customers frequent, how often they eat out, and how much they spend on an average visit need to be answered. Data from these questions can then be summarized and used to determine what products and services customers might need, how much they will be willing to spend, and how frequently they will repeat their business.

The Competition

It is important for entrepreneurs to be familiar with the competition. The **competition** includes any businesses that directly compete for an organization's customers. By understanding who the competitors are, what products and services they offer, how much they charge, and what makes them either successful or unsuccessful, entrepreneurs can make their own marketing efforts more effective. The most effective way to know the competition is to visit and analyze them. Figure 19–1 (page 260) shows a competition survey form that can be

COMPETITION SURVEY SHEET

Restaurant or Foodservice Outlet ____________
Address ____________
Date ____________
Completed by ____________

1. LOCATION
   - A. *Where in the community is the restaurant located?* ____________
   - B. *Access from major roads* ____________
   - C. *Parking* ____________
   - D. *Sign Visible* ____________ *lighted* ____________
   - E. *Access from public transportation* ____________
   - F. *Availability of public transportation* ____________
   - G. *Location features and liabilities* ____________

2. PHYSICAL APPEARANCE
   - A. *Architectural style* ____________
   - B. *Outstanding features* ____________
   - C. *General exterior condition* ____________
   - D. *General interior condition* ____________
   - E. *Types of dining areas* ____________

3. SERVICE FEATURES
   - A. *Days open (weekly)* ____________ *(yearly)* ____________
   - B. *Hours open (daily)* ____________ *(weekly)* ____________
   - C. *Seating capacity* ____________
   - D. *Turnover (per dining area and meal service)* ____________
   - E. *Average check (per meal)* ____________ *(overall)* ____________
   - F. *General cuisine* ____________
   - G. *Meals provided (breakfast)* ____________ *(lunch)* ____________ *(dinner)* ____________ *(brunch)* ____________ *(late supper)* ____________ *(other)* ____________
   - H. *Beverage services (lounge)* ____________ *(beer and wine only)* ____________ *(bar)* ____________ *(full beverage list)* ____________ *(wine list)* ____________ *(attached)* ____________
   - I. *Service (good)* ____________ *(mediocre)* ____________ *(poor)* ____________
   - J. *Additional facilities* ____________
   - K. *Entertainment* ____________
   - L. *Community acceptance* ____________
   - M. *Apparent problems* ____________
   - N. *Menu attached (yes)* ____________ *(no)* ____________
   - O. *Floor layout (provide description or sketch)* ____________

FIGURE 19–1 Competition Survey[2]

used to collect the information needed to determine how other food service establishments operate.

IDENTIFYING AND CREATING NEEDS

A main source of entrepreneurial ideas is the process of identifying needs that are not being met. Entrepreneurs can also create needs by recognizing and promoting desirable products and services of which customers are unaware.

Identifying a Need

A need can be identified in many ways. One way is through customer comment cards, such as the one shown in Figure 19–2. These cards, reviewed on a daily basis, keep management aware of customer com-

[2] Nancy A. Scanlon, *Marketing by Menu*, 1st ed. (New York: Van Nostrand Reinhold, 1985), 28.

Joe Gentile's SEAFOOD SHANTY

Customer Comments

MANAGER'S INITIALS

1. Did you enjoy your visit? Yes___ No___
2. How was your food? Excellent___ Good___ Average___ Poor___
   was the hot food served? Hot___ Warm___ Cold___
   was the cold food served? Cold___ Warm___ Hot___
   were the portions? Large___ Average___ Small___
3. Was your service satisfactory? Great___ Good___ Average___ Poor___
4. Would you recommend us to others? Yes___ No___

☐ Lunch ☐ Dinner

Suggestions or Comments: ___

Store #___

Server Name/ID___

Table #___

Please fill in information below to be eligible for weekly drawing.

Name ___

Address ___

City ___ State ___ Zip ___

Phone No. Day ___ Evening ___ Date ___

comment card

7/88

FIGURE 19–2 Customer Comment Card

plaints as well as requests for changes in current menu items or for new products and services.

A cookie company that distributes nationally, for example, filled a need by identifying restaurant customers who wanted consistently high-quality cookies. Originally a retail baking company that distributed exclusively through supermarkets, this company expanded its target market to include full-service food service operations. Simply by changing the packaging of its retail line (to accomodate bulk amounts of cookies) and identifying the most popular varieties for restaurant use, the company developed a successful merchandising program for a new target audience.

Creating a Need

Creating a need means identifying a product or service that could fulfill a need if customers were made aware of its existence. For instance, the Handy Soft Shell Crawfish Company in Louisiana (see Figure 19–3 on page 262) processes and sells soft shell crawfish as an alternative to shrimp, crab, or lobster. This shellfish resembles a lobster but ranges in size from three to five inches in length.

Promoting the item as a substitute for other, more expensive types of shellfish in appetizers and entrées, the company developed an effective marketing program and merchandising campaign. As one of their marketing techniques, they provided recipes and photographs to chefs and restaurants to promote the use of crawfish.

FOOD SERVICE MARKETING TRENDS

Identifying a need is the most important part of developing new products and services. Throughout this book,

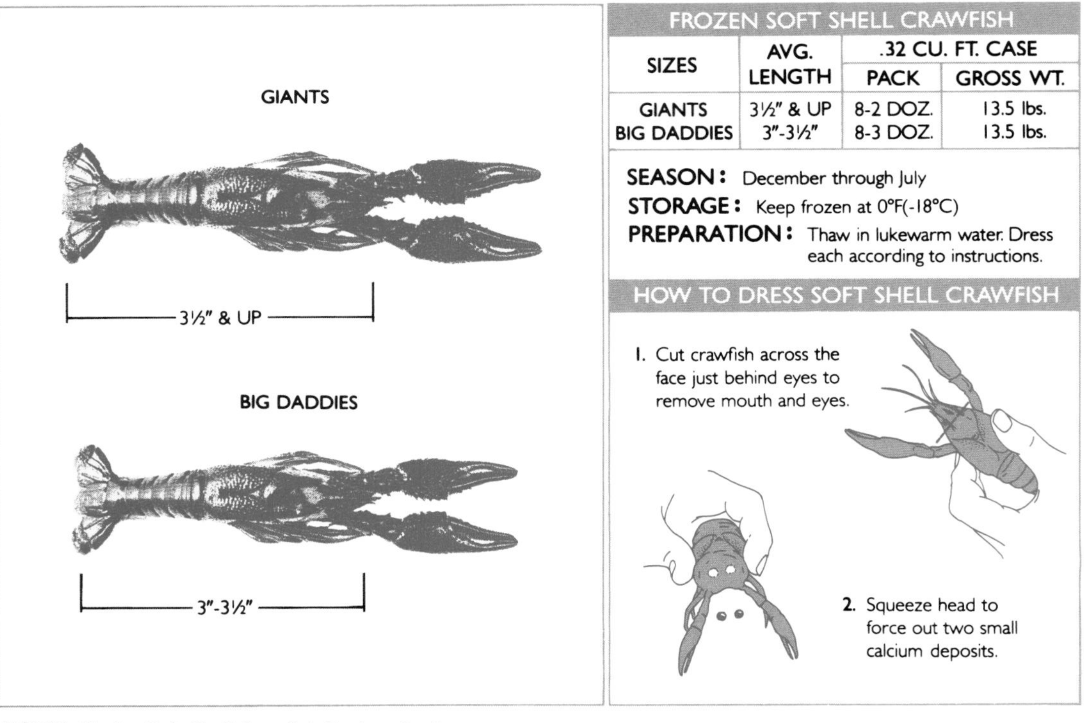

SIZES	AVG. LENGTH	.32 CU. FT. CASE	
		PACK	GROSS WT.
GIANTS	3½" & UP	8-2 DOZ.	13.5 lbs.
BIG DADDIES	3"-3½"	8-3 DOZ.	13.5 lbs.

FIGURE 19–3 Soft Shell Crawfish Product Card

ILLUSTRATION 19-2 As one of their marketing techniques, the Handy Soft Shell Crawfish Company provided recipes and photographs to chefs and restaurants to promote the use of crawfish.

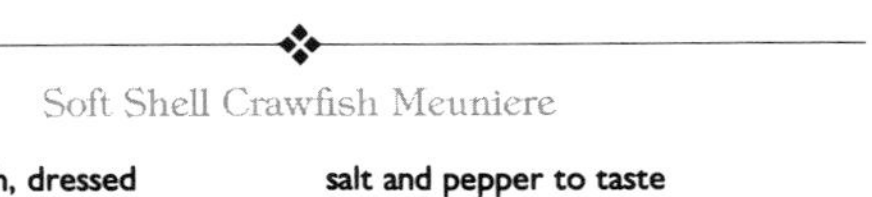

Soft Shell Crawfish Meuniere

8 soft shell crawfish, dressed
4 strawberries, puree (optional)
4 fleurons
2 cups water
8 oz. butter
salt and pepper to taste
2 lemons – squeezed for juice
½ cup demiglace
½ cup white wine
1 drop red wine vinegar

In a skillet, add water, vinegar, juice from one lemon, and salt. Delicately place crawfish in skillet. Cover with napkin and bring to gentle boil. Turn off and let sit for ten minutes.

Remove crawfish, and place on a dry towel.

Sauce:

In a separate pot, reduce demiglace with white wine and juice from one lemon for 8 to 10 minutes. Little by little, add solid butter and strawberry puree (optional). Stir until butter is melted, then add salt and pepper.

Display sauce in bottom of plate. Place soft shell crawfish in plate and garnish with fleurons. (2 Servings)

you have learned about ways in which the work habits and leisure life-style of the American public has changed. These changes lead to different needs, which in turn lead to new trends. A **trend** is a current style or preference. There are trends in many areas of American life, including the clothes we wear and the food we eat.

Marketing research has identified a number of major trends in the food service industry that are being directed by changing customer needs. Some of these trends are well established, with marketing and merchandising programs already in place. Others are just beginning to make an impact. All of these trends function as part of the 1990s food service industry. The major food trends today fall into three categories: convenience, dining, and cuisine.

Convenience Trends

Convenience trends are a direct result of the changing American work force. More that 55 percent of American women are now employed outside the home. In most families, both the woman and the man work outside the home. The number of single-parent families is increasing, largely due to divorce. These factors have significantly limited the amount of time available for at-home food preparation and food-related shopping. The result is a demand for food services that can provide fully or partially prepared food items, to be taken out and

ILLUSTRATION 19-3 Convenience trends, such as this drive-through window at a fast-food restaurant, are a direct result of the changing American work force.

eaten at home. Such food services are available from a wide range of restaurant operations and are receiving a positive response from the general public.

Pricing becomes an important part of the marketing cycle for convenience products. These items have a higher production cost because of the increased operational and labor expenses involved with preparing them. Therefore, convenience items have a higher selling price than unprepared food products. Customers, however, will not pay the price unless they feel that the value of the product is worth the cost. How much buyers feel that a product or service is worth is called the **perceived value.**

Dining Trends

Dining trends are focused on the ways in which people want to eat their meals. Food industry forecasters claim that a major shift in eating patterns is underway. Americans are eating smaller amounts of food, but more frequently. The custom of three major meals per day—breakfast, lunch, and dinner—is giving way to a pattern of five or six small meals eaten throughout the day. In some cases, a main meal is never eaten, while in other cases there is a single main meal. The family dining practice of eating breakfast and dinner together has changed as dramatically as the American lifestyle. Family members often eat whenever they are able to or in shifts, controlled by complicated schedules of work, recreation, education, and transportation. Two-income families often have overlapping work schedules, with one member supervising a meal while the other is at work.

Single-serving food products allow individuals to have a choice of menu items that can be cooked in microwave ovens. Nutritional snack items are also becoming more popular. Statistics on teenagers' spending patterns in convenience food stores show a marked increase in the amount of money being spent on snack items in the last five years. In restaurants the impact of these new eating patterns is seen in the increase of fast-food restaurants offering more wide-ranging menu selections. Salad bar and hot-food bar selections as well as a variety of sandwiches all fall into the category of small-meal items. In addition, full-service restaurants are offering a larger selection of appetizers. Customers can order two or three appetizers, a salad, and a dessert, completing their meals without ever eating an entrée.

One of the most significant recent dining trends has been an increase in nutritional and general health awareness. Every year, it seems, a new dietary concern appears on the scene. *Low-sodium, high-calcium,* and *low-cholesterol* foods and diets are just a few of the buzzwords related to healthy eating. Today's consumers are attracted to fresh, natural food items and preparation styles that retain vitamin and mineral contents.

Cuisine Trends

American cuisine has its roots in the 13 major native American cultures. These individual cultures, scattered across the United States, have developed many of the regional dishes that are now a traditional part of American cuisine. Corn-based products, such as the johnny cakes of the Wampanoag Tribe of Rhode Island and the corn meal shells of the Hopi Tribe of Arizona, have provided the base for all the native American cuisines.

American cuisine in the 1990s has blended meatloaf, mashed potatoes, and apple pie with the Cajun spices of the Louisiana bayous. Southwestern cuisine has brought together the influences of Mexico and California. Northwestern and Midwestern cuisines have combined freshwater fish, such as salmon and trout, with dairy products. Throughout the Grain Belt of the Great Plains, American cuisine has gathered the plentifulness of

ILLUSTRATION 19-4 Nutritional snack items are becoming more popular.

America's breadbasket. The South has contributed the plantation traditions of ham, beaten biscuits, grits, rice, and turnip greens, along with white layer cake and homemade ice cream.

The Mid-Atlantic states have given crab, oyster, and saltwater fish recipes to American cuisine. These dishes have been combined with the culinary influences of the immigrant populations that settled in that part of the country. Northeastern cuisine has combined the traditions of colonial America with the regional foods of New England. From pasta to pretzels to sauerkraut, the food items of major European cuisines have also become established elements of American cuisine as it heads into the twenty-first century.

ILLUSTRATION 19-5 American cuisine in the 1990s has blended many foods with the Cajun spices of the Louisiana bayous.

American chefs today are creating new and interesting variations to enhance the core of traditional American menu items offered in restaurants around the United States. This trend is often called **American nouvelle cuisine,** meaning the new American cuisine. Creative women and men continue to develop the American cuisine of the twenty-first century, building on two hundred years of traditions. American cuisine promises to undergo even more exciting changes to meet new customer needs and demands, providing entrepreneurs with diverse opportunities in the food service industry.

OPPORTUNITIES FOR ENTREPRENEURIAL ENTERPRISES

In food marketing, a great deal of emphasis is being placed on developing single-serving meals and products, especially for children. Many new items that are microwaveable and that children respond to are appearing on supermarket shelves. Microwave oven manufacturers estimate that four of every five households have a microwave oven.

U.S. census figures indicate that the population group aged 5 to 13 years old will maintain its size, between 30 and 34 million individuals, until 1990. Estimates on the growing number of children who are responsible for making many of their own snacks and meals are reinforced by work-force statistics. These figures indicate that 25 percent of all families have two-parent incomes, and another 25 percent are headed by a single parent.[3] With parents working outside of the home, convenience and safety become major concerns in food preparation for children.

All of these factors create opportunities for entrepreneurs willing to address the need for single-portion snacks and meals that children can handle safely. Customers want these products to be available in both perishable and nonperishable packaging. Those items that require heating should be packaged for microwave cooking. Prices on these items must respond to the needs of working families, including both single-parent and dual-parent incomes. To appeal to health-conscious consumers, these items should be nutritious, composed of natural food items, and contain as few chemical additives and preservatives as possible.

Major food manufacturers have identified this customer need and are currently developing lines of single-portion, microwaveable meals and snacks for children. Hormel is testing a line called Kid's Kitchen. My Own Meals (MOM) is distributing a brightly packaged line of frozen meals. Campbell's, Oscar Meyer, and Chef-Boy-Ar-Dee are marketing similar microwave versions.

ILLUSTRATION 19-6 With parents working outside of the home, convenience and safety become major concerns in food preparation for children.

Older Americans are on the other end of the age spectrum. Their numbers, and their needs, will continue to grow as the baby boomers reach retirement age. This group demands the same type of food packaging and pricing as that demanded by the children of working parents, but they require a different line of food items. These food products must respond to the low-sodium, low-cholesterol, and low-calorie diets followed by many older Americans. Both these markets—children of working parents and older Americans—present opportunities for applying the creative imitation and innovation theories of entrepreneurship.

[3] Wally and Marian Wood, "Kids Meals Go Big-Time," *Food and Beverage Marketing* (May 1989): 34.

CAREER PROFILE
EUGENE J. GAGLIARDI: OWNER, FOOD FANTASIES, INC., WILMINGTON, DELAWARE

DESIGNER
FOODS, INC.

Working with meat was a family affair at the butcher shop of Gene Gagliardi's father. From an early age, Mr. Gagliardi spent time sorting meat trimmings and learning how to butcher. He always tried to find better and faster ways to butcher and trim meat cuts.

One result of his experiments was the frozen beef product Steak-Umms. Mr. Gagliardi next directed his interest toward developing Australian lamb for the American meat market. His company's subsidiary, Designer Foods, Inc., ships gourmet cuts of lamb to restaurants throughout the world.

His latest venture in entrepreneurship has been to produce a boneless spare rib product called Spare the Ribs. To produce this product, rib meat is removed from the bone, thinly sliced, texturized, and blended in specially designed mixing machines. It is then extruded into strips through a manufacturing process and formed into sections of six to eight "ribs." Currently being featured in major fast-food chains around the country, this product promises to be another entrepreneurial success for Mr. Gagliardi.

SUMMARY

Entrepreneurship is the basis of the American food service industry. By applying the theories of entrepreneurship, including leadership, creative imitation, and innovation, many businesses have developed. Effective entrepreneurs have knowingly (or unknowingly) applied the three keys to success: identifying the target market, developing a customer profile, and knowing the competition. Before entrepreneurs can turn their ideas into successful businesses, however, they must identify or create a need.

The many changes in American lifestyles continue to create opportunities for entrepreneurial food-related enterprises. Dining and cuisine trends are evolving due to the American public's growing interest in and need for convenient and healthy food products and services.

Those individuals and companies who accurately identify the needs of the 1990s customer have the opportunity to introduce exciting and creative products and services to the marketplace. They must, however, be willing to risk the failure of their ideas. In addition, they must have the energy and stamina required to make any business a success.

QUESTIONS

1. What does the term *entrepreneur* mean?
2. Besides the basic principles of business management, what else must entrepreneurs apply to their ideas?
3. What are the three theories of entrepreneurship? Which two are most commonly applied to entrepreneurial food service ventures?
4. What is the process for identifying and creating a new product or service?
5. What are the entrepreneurial keys to success?
6. How has McDonald's identified and successfully responded to customer needs?
7. What does the term *perceived value* mean?
8. What factors must an entrepreneur know about her or his customers in order to have a successful enterprise?
9. What do the terms *target market, customer profile,* and *competition* mean?
10. What three trends influencing food service marketing are discussed in this chapter? Select and explain one of these trends in detail.

ACTIVITIES

Activity One As a class activity, conduct a survey of food service establishments in your community to determine how and why they were opened. Identify which ones are the result of local entrepreneurial efforts rather than part of a national chain.

Activity Two Think up an idea you could develop into an entrepreneurial business. Write down the idea in paragraph form. Identify the need that you will fill in your community. Using the guidelines in this chapter, identify your target market and your competition, and then outline a customer profile.

Activity Three Find an entrepreneur in your community who has developed a food-related idea into a successful business. Arrange for an interview with this person. Using the information in this chapter, ask questions about how she or he began in business. In what ways did this person use the theories or guidelines discussed in this chapter? Present your findings in an oral report to your class.

20 ORGANIZATIONAL RECORD KEEPING

VOCABULARY

purchase order
purchase requisition
order sheet
invoice
reconciled
inventory
production sheet
daily food cost sheet
sales mix record
leaders
losers
sales history form
daily server report
daily server sales report
accounts payable
accounts receivable

OBJECTIVES

After studying this chapter you should be able to do the following:

- Explain the relationship of organizational record keeping to the success of a food service business
- Be able to identify the four major areas of a restaurant operation that create and direct organizational records
- Chart the flow of record keeping from the point of supply to inventory
- Create an outline of the organizational records necessary for the success of a food service operation, including the different locations in which these records are typically generated
- Discuss the server's role in accurately recording information on the guest check and the relationship of these records to the record-keeping base

Organizational record keeping is the basis for the successful functioning of any food service business. Unless records are consistently and accurately maintained, the daily operations of a restaurant will fail. In today's highly competitive food service industry, the margin for error is slim. Accurate records of purchases, food costs, production costs, inventories, and menu sales must be made in order to effectively calculate menu prices and food cost percentages.

Planning is an important part of successful restaurant management and is based on a series of production and service records. Food service record keeping involves four major areas of an operation: purchasing, production, service, and personnel. Organizational records are created in and directed from four different locations: the purchasing agent or department, the kitchen, the dining room, and the manager's office.

Although examples of a number of forms are illustrated in this chapter, it should be noted that fully computerized food service operations use few forms. Computer programs can handle almost all of the production and service records for an operation. Personnel-related forms, however, are generally recorded by hand. This chapter discusses the record-keeping procedures necessary for the operation of a food service business. The formats for many computer-generated records are covered in Chapter 21.

PURCHASING RECORDS

Purchasing records follow the path of purchased goods as they go through a restaurant. The first form used in this process is the purchase order. Others include the order sheet, invoice, and inventory form.

The Purchase Order

A **purchase order** (or a **purchase requisition**) is filled out by department heads, such as a dining room hostess or host, chef, head bartender, or man-

ILLUSTRATION 20-1 Purchasing records, such as the forms this manager is studying, follow the path of purchased goods as they go through a restaurant.

ager, to request needed goods or services. The chef, for example, orders food supplies to produce menu items, and the hostess requests paper supplies for the dining room. Nonalcoholic and alcoholic beverages for bar service are requisitioned by the head bartender, and the manager fills out purchase orders for equipment. If an operation is well organized, each item is identified by an inventory number. This number is part of an inventory system that allows for an accurate accounting of the supplies on hand. If the purchase order form produces a carbon duplicate or triplicate, a copy may be returned with the order when it is filled.

The purchasing agent, or whoever is responsible for ordering, receives purchase orders and issues supplies as they are available in inventory. Items that are not in inventory are placed on a master order sheet. Perishable food items are always ordered as close as possible to the date of their use.

An organized chef and purchasing agent plan two to three weeks ahead for major menu items, such as meat and poultry. Wholesale food prices fluctuate daily, and knowing what is needed in advance can help ensure cost-effective purchasing.

The Order Sheet

When the purchasing agent has all of the orders on hand and has reviewed the required food items with the chef, an order sheet is filled out. The **order sheet** lists all of the items that have been requested but are not in inventory, broken down by category, such as paper supplies or food items. The size of the unit in which an item is generally purchased is also listed, along with the unit price. When ordering by telephone, the purchasing agent uses the order sheet while requesting goods from purveyors and suppliers, recording the current prices, the availability of items, and the order as it is placed. This information can then be transferred into a computer system, if one is available.

The Invoice

When items are received they are accompanied by an invoice, or bill. The **invoice** is an itemized list of goods shipped, specifying their prices and the terms of the sale (see Figure 20–1). The invoice should be signed at the time of delivery to indicate that the goods were delivered as stated. Ideally, all food products should be inspected for their condition and quality. The invoice is matched with the purchase order or the order sheet to make sure that the number of items delivered and prices charged are the same as those originally placed and quoted. When the invoice has been **reconciled** (checked against the order), it is then signed by the purchasing department and sent onto the manager's office for payment.

INVOICE

WINEBERRY FARM
A-1014
641 Lancaster Pike
Frazer, Pa. 19355
215-647-6903

NO. ____
DATE ____
YOUR ORDER NO. ____
OUR ORDER NO. ____

SOLD TO: ____

SHIP TO: ____

FOB	TERMS	DATE SHIPPED	SHIPPED VIA	SALESMAN

	ORDERED	SHIPPED	DESCRIPTION	PRICE	PER	AMOUNT
1						
2						
3						
4						
5						
6						
7						
8						
9						
10						
11						
12						
13						
14						
15						
16						
17						
18						
19						
20						

FIGURE 20–1 An Invoice

The Inventory Form

An **inventory** is a list of the supplies and food items on hand (see Figure 20–2 on page 274 for a sample). A central inventory sheet will list all of the items that are used in a business and assign an inventory identification number to each of them. This number is then used on inventory forms, purchase orders, order sheets, and all general accounting records. Inventory for food service operations should be done on a weekly basis, if not daily.

Because food products are so perishable, it is necessary for both the purchasing agent and the kitchen staff to help keep food waste at a minimum. Having an accurate and up-to-date inventory accomplishes two major functions. It allows the purchasing agent to order only what is needed from suppliers and facilities, using up the current inventory. It also enables the chef to change the food production schedule to use up food items that have not been processed for one reason or another. If the amount on hand is known, the purchasing agent can take advantage of special bargains by volume purchasing. Proper inventory records can thus result in lower food costs and higher profits.

INVENTORY ______________________ PAGE ______

SHEET NO. ______________________ PRICED BY ______

CALLED BY ______ DEPARTMENT ______ EXTENDED BY ______

ENTERED BY ______ LOCATION ______ EXAMINED BY ______

Check	Quantity	Description	✓	Price	Unit	Extensions
		Amount forward				

FIGURE 20–2 An Inventory Form

PRODUCTION RECORDS

Production records are generated by a chef's office to forecast the amount of food that is needed, to record the number of menu items that are actually produced and served, and to calculate food costs and food cost percentages. These records also serve as the basis for future planning. Production records include the production sheet, daily food cost sheet, sales mix record, and sales history form.

The Production Sheet

A **production sheet** is a form used to forecast, or predict, the number of servings of each menu item that will be produced for a given meal service on a given day. This form is the basis on which a chef requisitions food supplies and schedules kitchen personnel.

Using information on the production sheet, an estimated food cost can be calculated based on the cost card for each menu item. Following the

NUTRITION NOTE KEEPING RECORDS TO MAINTAIN FOOD QUALITY

Planning in a successful food service operation involves being able to accurately forecast the amount of food that is needed over a given period of time. Many record-keeping forms, especially production sheets, daily food cost sheets, and sales mix records, help management to predict food needs and costs. Accurate records allow the chef and purchasing agent to order food supplies well in advance and to take advantage of seasonal availability and better prices. In addition, purchases of perishable fruits and vegetables can be made as close as possible to the dates they are used. The fresher the produce, the higher its nutritional content. As produce ages, many of its valuable vitamins and minerals are depleted.

A good inventory system helps to eliminate food waste. Storage should be based on the maxim FIFO, "first in, first out." This principle means that food items already in storage must be used before the new items that are purchased to replace them. This rule is particularly important for fresh and frozen food items. Frozen foods have a limited shelf life in a freezer. After they have been frozen for a period of time, food items begin to deteriorate and do not have the nutritional value or the taste quality that they had when originally frozen.

---

service, the number of each item served is recorded on the production sheet. The number of leftover portions are also recorded to calculate overproduction and waste. Some food products, such as large cuts of meat and poultry as well as fresh fruits and vegetables, can be reconstituted or made into other products. Overproduction, however, can often result in waste, which is costly and affects food cost percentages. It is important that management be aware of how much waste has occurred and the cost of that waste to the operation.

Kitchen planning for menu items should be based on the sales history of the restaurant and the forecasts for daily and monthly business. Food costs and overall profits can be controlled more precisely when the number of items produced is the same as the number of items served.

The Daily Food Cost Sheet

A **daily food cost sheet** is a record of the total cost of purchase orders or purchase requisitions balanced against the total sales for a day. This form produces both a stated food cost and a food cost percentage. The food cost formula given in Chapter 12 (FC ÷ SP = FC%, or food cost ÷ selling price = food cost percentage) is used to calculate the total food cost percentage. The daily food cost sheet, such as the one shown in Figure 20–3, acts as an ongoing record of daily and monthly food costs for an operation.

The Sales Mix Record

A **sales mix record** shows the amount of each menu item sold over a specified period of time, usually a month. This record provides a basis for forecasting production schedules and for planning volume purchases.

The sales mix record allows management to look at the overall sales for the restaurant, item by item, and to see changes in customer preference. Menu items that are selling poorly can be identified, and their position on the menu can be evaluated. Using past sales mix records, the item's history can be reviewed to see if the drop in sales is seasonal (because of the type of food item), caused by production problems, or a result of poor server sales techniques. If customers' perceptions of the menu item has changed because of outside influences, such as publicity about the relative health benefits of different types of food, management may have to remove the item for a period of time or change its name and presentation.

FIGURE 20–3 A Daily Food Cost Sheet

DAILY FOOD COST REPORT

Location: ____________________

					Monthly Totals		
Date	Day	Total Cost	Total Sales	Food Cost %	Total Cost	Total Sales	Food Cost %

Every item on a menu is not intended to sell well. Depending on the style of an operation, customers expect to see certain types of food items on the menu, and management knows in advance that some of these items will be more popular than others. The cuisine and general theme of an operation will determine which menu items are the most popular. Items that have a high sales volume are called **leaders,** and items that have a low sales volume are called **losers.**

A full-service restaurant, for example, usually offers an appetizer menu that includes shrimp, escargot, crab, and fruit. The leader on this menu is generally the shrimp, followed by the crab. Escargot and fruit sales fluctuate on a seasonal basis. The restaurant managers study this sales mix and see that they can buy volume quantities of whole shrimp and crab. Because the

MATH SCREEN

Fill in the missing figures for the dates provided on the following daily food cost sheet. Use the food cost formula FC ÷ SP = FC% (food cost ÷ selling price = food cost percentage) to calculate the food cost percentages. Round up to the next highest number when calculating percentages.

DAILY FOOD COST REPORT

Location: AUDREY'S

Date	Day	Total Cost	Total Sales	Food Cost %	Monthly Totals: Total Cost	Monthly Totals: Total Sales	Monthly Totals: Food Cost %
FEB. 15	MON.	254.16	882.75	29			
FEB. 16	TUE.	312.86	910.30		567.02	1,793.05	
FEB. 17	WED.	301.20	792.45				
FEB. 18	THUR.	295.18	922.20				
FEB. 19	FRI.	452.13	1,006.10				

ILLUSTRATION 20-2 Items that have a high sales volume are called leaders. Shrimp, for example, is generally the leader on an appetizer menu.

escargot is canned and has a considerable shelf life, it can be purchased in advance without worrying about spoilage. The fruit can, with careful planning, be reconstituted somewhere else on the menu if it does not sell as an appetizer. The sales mix record allows management to monitor the sales of each of these menu items.

The Sales History Form

A **sales history form** is the daily journal of a restaurant (see Figure 20–4). On this form, managers record special events or problems that may have affected their business on that particular day. The number of reservations, the total number of customers served at each meal service, and the total revenue are also noted. The manager on duty can make comments on the sales history form that might help to plan for that particular day one year later.

FIGURE 20–4 A Sales History Form

SALES HISTORY

Location:

Date: Day:

Weather conditions:

Number served:

Private parties:

Problems and comments:

Manager on duty: Initial:

SERVICE AND PERSONNEL RECORDS

A server's position as the main salesperson for a food service operation requires that a record of her or his effectiveness (that is, number of sales) be kept. The forms used for this purpose include the guest check, daily server report, and daily sales report.

The Guest Check

A server handles one of the most important record forms in a food service operation: the guest check. (See Figure 20–5 for a sample.) This person must accurately record the number of customers served, the menu and beverage items ordered, and the prices charged. The guest check is used as the basis for information on the production sheet, the daily food cost report, and the sales history form. If an operation is computerized, then the server becomes responsible for entering correct information into the computer through the register or the keyboard.

FIGURE 20–5 A Guest Check

CHECK NO. 140553	DATE	TABLE NO.	SERVER	NO. PERSONS

1	
2	
3	
4	
5	
6	
7	
8	
9	
10	
11	
12	
13	
14	
15	
16	
17 FOOD	
18 BEVERAGE	
19 TAX	
20 TOTAL	
21	

CHECK NO. 140553	DATE	NO. PERSONS	AMOUNT OF CHECK

The Daily Server Report

A **daily server report** records all of a server's transactions for a day by transferring the information from guest checks onto a report form. The number of guests served, total food sales, total beverage sales, taxes, and tips are listed. In addition, the type of sales, whether cash or charge, is noted.

The Daily Server Sales Report

A **daily server sales report** (see Figure 20–6 on page 280) records the transactions for all of the servers working in a restaurant on a particular day. This report allows management to review the sales totals for all servers. The information for each server is transferred from the daily server reports.

THE ACCOUNTING DEPARTMENT'S ROLE

The operational records from the purchasing, production, service, and per-

Location:

Date:

DAILY SERVER SALES REPORT

Server Name	Number Served	Total Food	Total Beverage	Sales Tax	Total Sales	Cash Sales	Charge Sales
Totals							

FIGURE 20–6 A Daily Server Sales Report

sonnel areas are gathered together as the basis for the total record-keeping system that is completed by the accounting department. **Accounts payable,** a record of bills that must be paid, and **accounts receivable,** a record of accounts that are owed to the operation, are transferred from the information on invoices and guest checks. Checks are then issued in payment. Sales receipts and expenses are recorded monthly in the general accounting ledger, and totals are calculated. This ledger gives an overall accounting picture that allows management to see all of the overhead, labor, and food costs of the operation. This data enables managers to determine and to monitor the margin of profit or loss.

CAREER PROFILE

RONALD N. GORODESKY: ASSOCIATE, LAVENTHOL & HORWATH, DIRECTOR OF FOOD AND BEVERAGE CONSULTING PRACTICE, PHILADELPHIA OFFICE

LH Laventhol & Horwath
Certified Public Accountants/Business Consultants

Source: Reprinted with permission of Laventhol & Horwath.

Ron Gorodesky is a member of the national accounting firm of Laventhol & Horwath. This firm has been a leader in advising hospitality-related businesses for over 50 years throughout the United States and the world.

Mr. Gorodesky became involved with the food service industry in the same way as many other high school students do—he took a job as a dishwasher at age 15. From bus person to line cook to waiter, he held a variety of restaurant positions during high school. After graduation, he entered Pennsylvania State University planning to major in business. The Hospitality/Restaurant Management Program at Penn State, however, promised immediate job prospects upon graduation as well as challenging career possibilities. Based on his job experiences in high school and his interest in the food service industry, Mr. Gorodesky decided to major in hospitality and restaurant management.

His first jobs following college graduation ranged from converting hotel properties as a task force member for Marriott Hotels to managing restaurants for Hospitality Investments, Inc. In 1984 he joined the Leisure Time Industries Division of Laventhol & Horwath.

Currently Mr. Gorodesky's job focuses on supervising the food and beverage consulting efforts of his firm's Philadelphia office. These efforts include developing feasibility studies, marketing surveys, and operational food and beverage controls for new and existing hotel and restaurant food service operations. Mr. Gorodesky attributes his ability to manage his office's consulting business to his wide range of experiences in the food service industry.

SUMMARY

Food service operations require organizational records to successfully function in the business world. Many people who enter food service management do not understand the importance of accounting systems and fail to carry out proper record-keeping procedures. Most restaurants fail from poor management systems, not from poor food or service. Managers are responsible for good record-keeping and accounting practices to ensure the success of a food service business.

Organizational records are created in four major areas of food service operations: purchasing, production, service, and personnel. An operation's staff needs to be familiar with the various forms used in these areas and have the skills required to accurately complete them.

Managers must take the time to train their staff in organizational record keeping and to hire good accountants to complete their financial transactions. Food service computer systems now help management keep records and generate reports that are much easier and faster to use.

QUESTIONS

1. Why is organizational record keeping necessary to the success of a food service operation?
2. What four major areas of a restaurant operation create and direct organizational records? What are the records that originate from these areas?
3. Define the term *inventory*. Why is an accurate inventory important to the profitability of a business?
4. What is the importance of the production sheet relative to other production records?
5. How is the food cost percentage calculated on the daily food cost sheet? What costing formula is used?
6. What kinds of information do the sales mix records provide?
7. What is the difference between the terms *losers* and *leaders* relative to the sales mix?
8. What role does the guest check play in providing information for organizational records? What is the server's role in accurately recording this information?
9. Why is it important to record the individual sales of each server?
10. What cost factors does the general accounting ledger of a food service operation allow management to review?

ACTIVITIES

Activity One Make a presentation to your class charting the flow of record keeping from the initial point of supply to inventory.

Activity Two With the help of your teacher, create an inventory system for all of the products and supplies in your classroom that are related to food service. Assign an inventory number and a cost to each. Using an inventory form such as the one illustrated in Figure 20–2 (page 274), post the inventory and calculate its value.

Activity Three Identify a food service operation in your community that is large enough to use all of the organizational records discussed in this chapter. Interview the manager to determine which forms the staff completes, how effectively they keep records, and what other record-keeping systems, if any, they use.

Activity Four Outline the organizational records necessary for the success of a food service operation. Begin by identifying the four major locations of a restaurant from which records are directed. Display the outline in your classroom.

21

COMPUTERIZED FOOD SERVICE FUNCTIONS

VOCABULARY

POS system
remote printers
electronic cash register
memory
keyboard
monitor
hard copy
roll printer
central processing unit (CPU)
modular computer system
revenue per cover
percentage of yield
exploded

OBJECTIVES

After studying this chapter you should be able to do the following:

- Realize the impact computers can have on the successful functioning of a food service operation
- Describe the variety of computer equipment available to the food service industry
- Outline the POS modular computer system and then discuss the advantages of this system for a full-service restaurant
- Compare the manual and the computerized techniques of organizational record keeping
- Chart the flow of data in a food service operation, from a computer-generated guest check to the appropriate report forms

The computer is no longer the restaurant management tool of the future—it is now the most important piece of new equipment in commercial kitchens. Most food service operations fail due to a lack of proper management rather than poor food service. The computer, which enables the fast, accurate management of important records, can play a key role in business success.

As discussed in Chapter 20, full-time, experienced personnel are needed to process the large number of records needed to successfully run a business. A restaurant manager who tries to assume the responsibilities of kitchen and dining room supervision as well as record keeping will soon be overwhelmed. Computer programs relieve management of many record-keeping functions by transferring information from guest checks directly into the program's file for each type of record, where it is stored until needed.

COMPUTER SYSTEMS

From fast-food to full-service businesses, the computer has become a vital part of day-to-day restaurant operations. Computer systems can be as simple as an electronic cash register, or as complicated as a **POS system** (a system that begins at the *point of sale*) with remote printers and an integrated food service management program. **Remote printers** are printers located in areas away from the keyboard, CPU (central processing unit), and monitor. Information can be selected to print at different stations where a printer is located. This chapter reviews the range of food service computers and the food service operations for which they are most appropriate.

Electronic Cash Registers

The simplest computer system used in food service operations is the **electronic cash register.** The principle function of an electronic cash register is to record retail sales. This type of register will issue customer receipts as well as sales records. The secondary function of an electronic cash register is to assist a salesperson in recording and completing a sale. From the time that a transaction is entered into a register until the time that change is made, the register provides backup information. It tells an operator how much change to issue based on the amount of cash offered to purchase the item. It also presents information digitally, which helps an operator process transactions quickly. An electronic register has a limited **memory,** or capacity for storing data. Data can be printed on register tape at the end of an operator's shift to be used for inventory control and for the salesperson's sales record.

Register Systems in Fast-Food Restaurants

Fast-food restaurant chains use another register system at the point of service. In this type of system, each register's **keyboard** (a bank of keys by which a machine is operated) is clearly

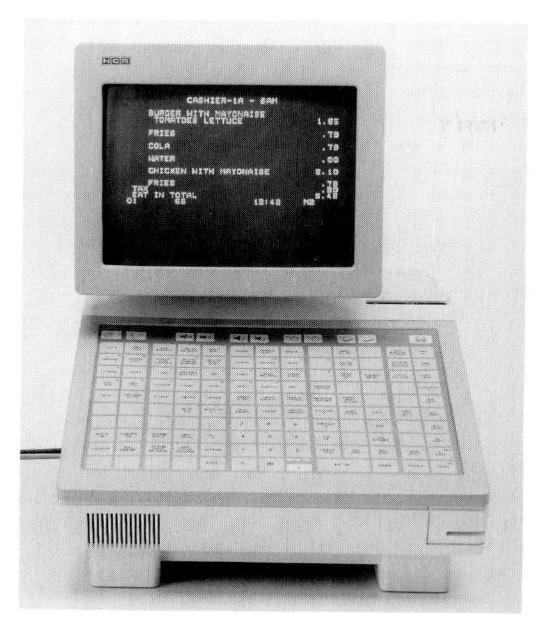

ILLUSTRATION 21-1 Fast-food restaurant chains use a register system with clearly marked keyboards and liquid crystal display panels at the point of service.

marked with the names and, often, the pictures of menu items on the keys. Because register attendants in fast-food operations are also servers, the registers have liquid crystal display panels to magnify lists of ordered items to increase efficiency and help eliminate service problems.

POS Systems in Full-Service Restaurants

Full-service restaurants often use POS systems. These systems require point-of-sale ordering panels in the dining room areas, remote printers for the bar and kitchen areas (where orders are printed), and complete software programs with both production and management capabilities.

The three pieces of equipment included in another type of POS system are a **monitor** (or screen), a cash drawer, and a keyboard. This equipment is at a server workstation along with a printer. The keyboard is touch sensitive and color coded to help servers distinguish between menu categories.

An additional type of ordering system, in the form of electronic server pads, is being offered by some computer companies. Remanco, a major manufacturer of restaurant computers, has a system that features an FM radio transmitter. The transmitter is carried by a server, who can transmit orders to remote printers in the kitchen and bar areas. The use of the hand-held pad keeps a server on the floor servicing customers. Large family style restaurants are ideally suited to this type of system because they need to provide speedy service.

Computerized Cash Registers

Restaurants that want record-keeping capabilities without monitors, POS ordering panels, or electronic server pads can use computerized cash registers. This system provides a keyboard, which can be personalized according to menu items, as well as record-keeping functions. Server sign-in and sign-out features, tip tracking, guest check

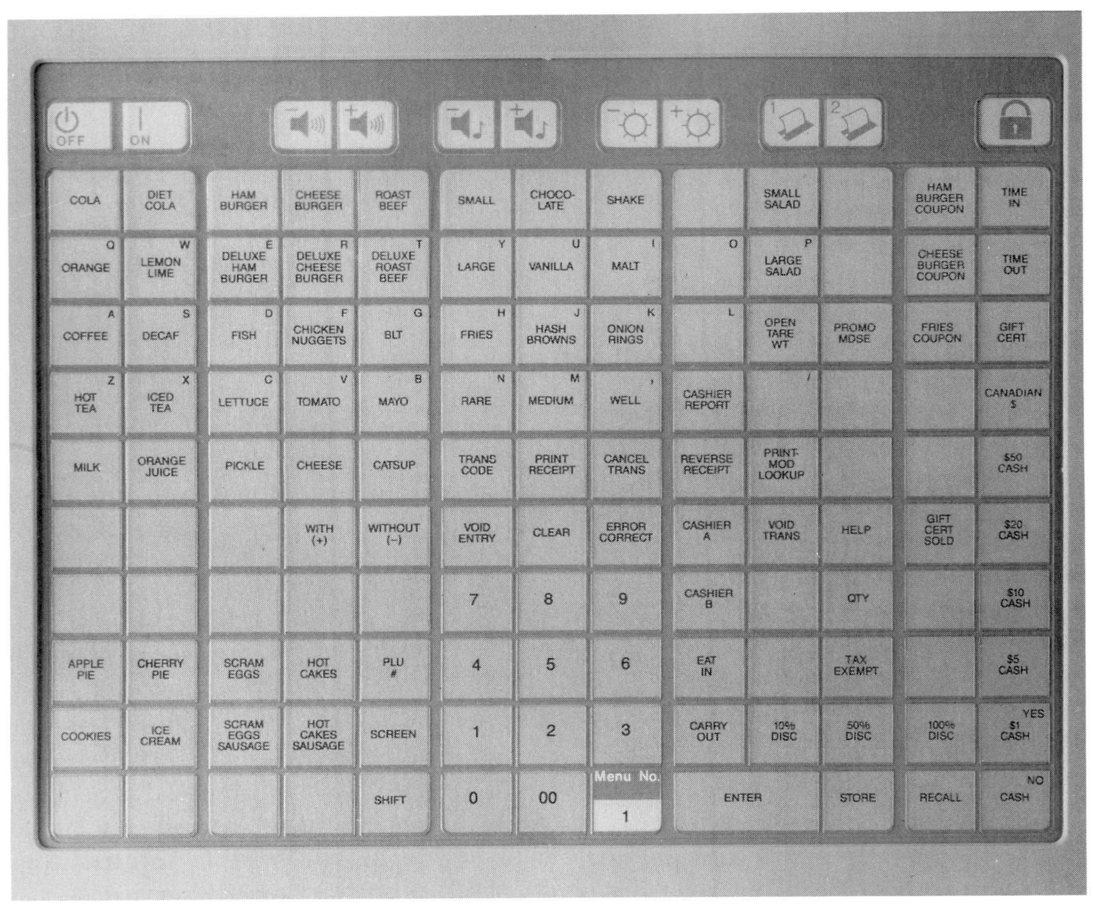

ILLUSTRATION 21-2 Register keyboards can be personalized according to menu items.

information, server sales records, sales reports, and a cash drawer are available with this equipment. Computerized cash registers have the capability to transfer information to a complete management system. They also have built-in printers that allow hard copy to be inserted. **Hard copy** refers to the printed record of a computer-generated transaction, such as the checks shown in Figure 21–1.

Printers

There are two common types of printers for front- and back-of-the-house restaurant use. A small roll printer is used as a remote printer at different stations in a restaurant, such as the bar and the kitchen. A **roll printer** prints on a roll of paper similar to the kind used in a cash register, and is appropriate for recording orders. A large printer is located at server work stations to print hard copy guest checks.

Back Office Equipment

The final part of a complete service system is the back office equipment. This equipment includes a more traditional computer keyboard, 80-column dot matrix printer, and a central processing unit. The **central processing unit (CPU)** stores data provided by the POS system and, with the keyboard and printer, generates all of the operational records discussed in Chapter 20 as well as spreadsheets and other accounting records.

Back office equipment combined with a POS system provides improved ordering accuracy for faster service and less waste. The CPU memory can store and control a number of complete menus as well as accomodate different methods of payment. It can continually update a guest check balance, printing out the total of the check on demand. This type of system speeds up cash flow accounting, and its server sign-on function makes payroll record keeping more accurate and simplifies tip tracking.

FIGURE 21–1 Hard Copy Checks

CHECK NO. 68191	DATE	TABLE NO.	SERVER	NO. PERSONS

CHECK NO. 68191	DATE	NO. PERSONS	AMOUNT OF CHECK

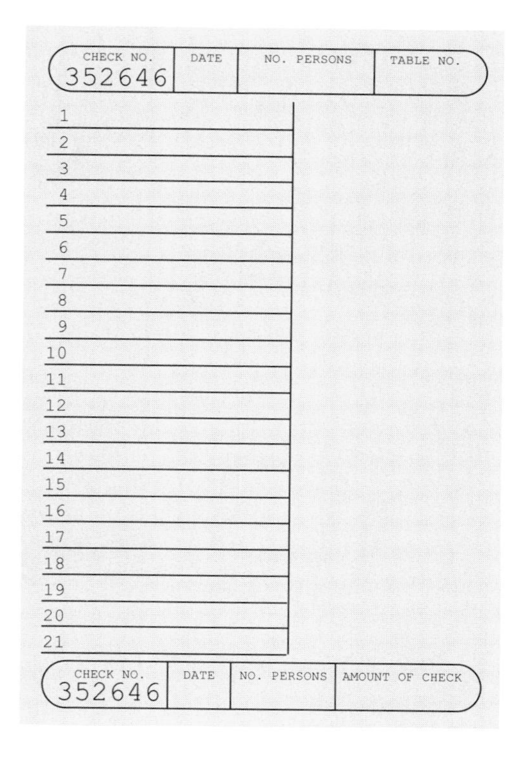

CHECK NO. 352646	DATE	NO. PERSONS	TABLE NO.

1
2
3
4
5
6
7
8
9
10
11
12
13
14
15
16
17
18
19
20
21

CHECK NO. 352646	DATE	NO. PERSONS	AMOUNT OF CHECK

Modular Computer Systems

In order to help you clearly understand how all of this equipment works within a food service operation, Figure 21–2 illustrates a modular computer system. A **modular computer system** consists of a number of separate stations that can operate independently of one another or in unison. In Figure 21–2, for example, customer orders are input at stations one and two and transferred to the bar and kitchen areas by the remote printer in

FIGURE 21–2 A Modular Computer System

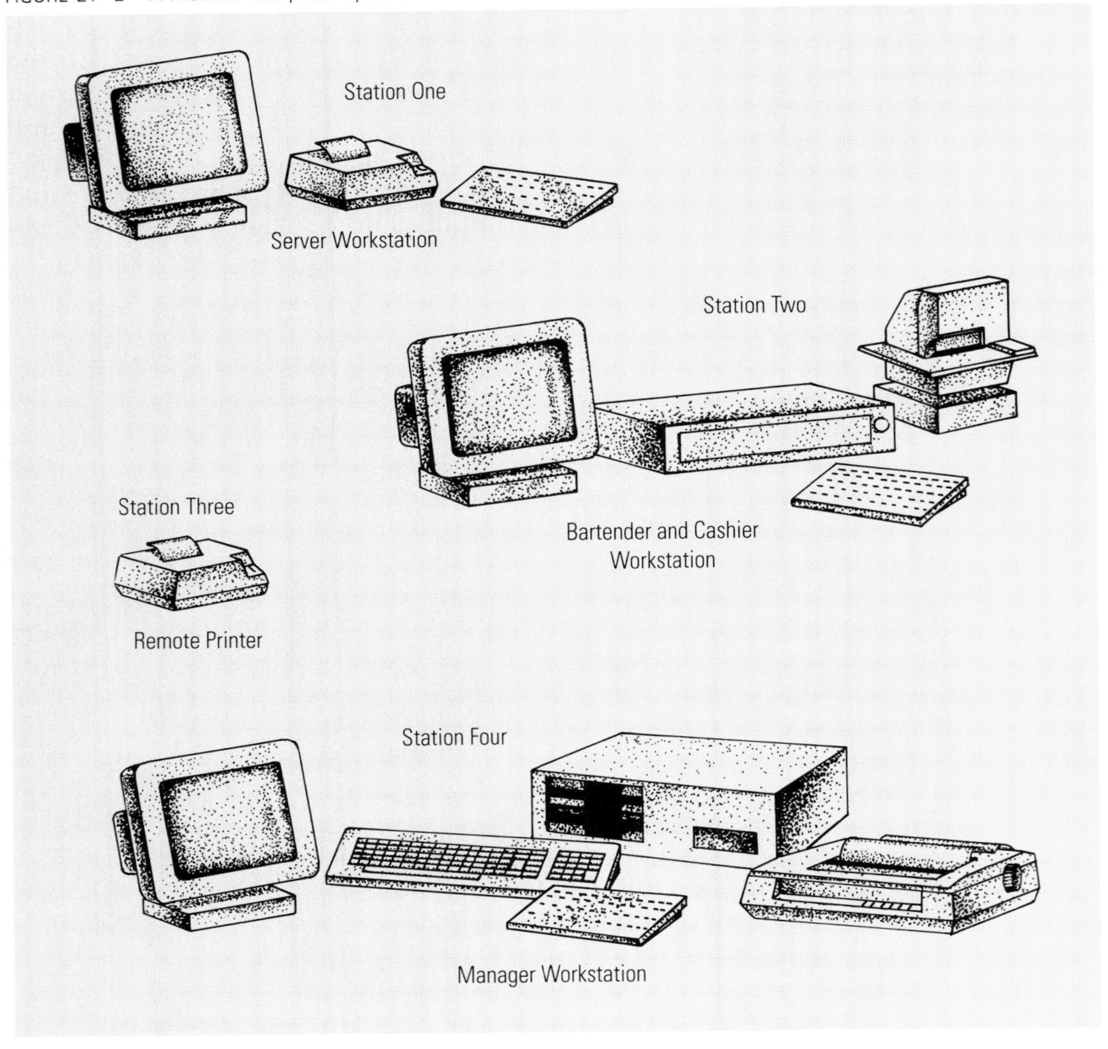

station three. Station three does not generate any information, only relays it. Station four receives data from stations one and two but does not transmit information to the remote printer in station three.

A restaurant operation may begin its investment in a computer system with a POS server workstation and a remote printer in the kitchen. At a later date, it may add a manager's workstation and a bartender or cashier station.

COMPUTERIZED ORGANIZATIONAL RECORD KEEPING

The computer systems discussed in the preceding sections provide a variety of record-keeping capabilities. The primary source of data for computer-based records is the guest check. The information on a check is relayed to the appropriate record forms when a transaction is closed out at a register and a guest has paid his or her bill. Figure 21–3 breaks down guest check information and identifies the different report forms that receive data.

Employee Sales Profile Reports

A daily server report (discussed in Chapter 20) reviews a server's activity for one day and must be filled in by hand. This information can be computerized by transferring sales information from guest checks to an employee sales profile report file in a computer, using the server's name as

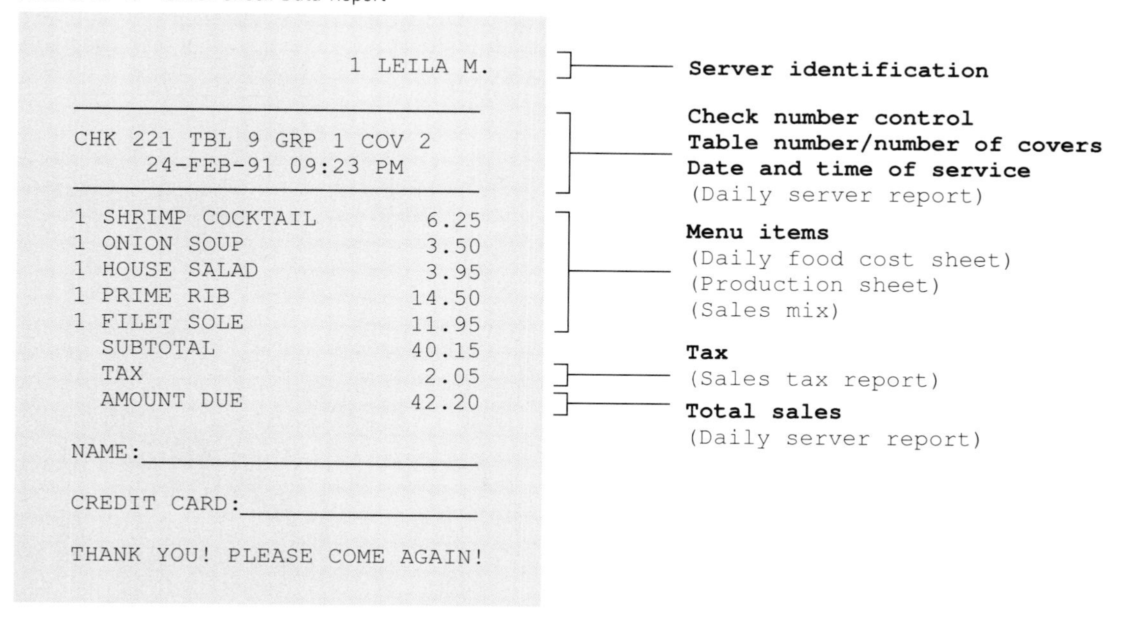

1 LEILA M.

CHK 221 TBL 9 GRP 1 COV 2
24-FEB-91 09:23 PM

1 SHRIMP COCKTAIL	6.25
1 ONION SOUP	3.50
1 HOUSE SALAD	3.95
1 PRIME RIB	14.50
1 FILET SOLE	11.95
SUBTOTAL	40.15
TAX	2.05
AMOUNT DUE	42.20

NAME:

CREDIT CARD:

THANK YOU! PLEASE COME AGAIN!

FIGURE 21–3 Guest Check Data Report

identification. Some operations will assign a code number to avoid any name duplication problems. This report records a server's total sales for each day, the number of customers served, the number of guest checks used, and also the number of individual tables served.

The report in Figure 21–4, for example, shows that 15 tables were served for the day. The server was assigned 5 tables and the total amount of time that customers spent at those 5 tables was 14 hours and 22 minutes. The average **revenue per cover,** or the amount of money spent by each customer, was $6.65. The average revenue per guest check was $18.10, and the average revenue per table was $65.15. Each table was turned three times, with an average revenue of $21.72 per table turn. A table turn refers to turning a table back into use after the guests seated at the table have left the restaurant. The report finishes by noting an average of 1.2 checks issued per table turn and an average of 3.27 customers per table turn. The average amount of time spent by guests at a table was 57 minutes.

FIGURE 21–4 Employee Sales Profile Report

EMPLOYEE SALES PROFILE REPORT MICROS 4700 VERSION 3.00 SYSTEM 02-SEP-91 05:52 PM Page 1
RANGE 5-14 C: PRIMARY DATABASE

EXAMINE POSTED 02-SEP-91 05:29 PM CURRENT RESET 0002 02-SEP-91 02:30 AM
TO-DATE RESET 0001 01-SEP-91 02:35 AM

11 Cosby, William RESTAURANT

	CURRENT TOTALS	TO-DATE TOTALS
Net Sales Revenue	325.76	684.44
Covers Count	49	104
Guest Check Count	18	42
Table Turn Count	15	35
Table Count	5	10
Table Dining Time	14:22	36:30
Revenue/Cover	6.65	6.58
Revenue/Check	18.10	16.30
Revenue/Table	65.15	68.44
Turns/Table	3.00	3.50
Revenue/Turn	21.72	19.56
Checks/Turn	1.20	1.20
Covers/Turn	3.27	2.97
Dining Time/Turn	0:57	1:03

ILLUSTRATION 21-3 An employee sales profile report shows the average amount of time spent by guests at a table.

The to-date (meaning month-to-date) totals in Figure 21–4 are two-day totals. The comparison between current and to-date totals allows management to assess a server's sales record on a day-to-day basis, according to the number of customers served, the total revenue per customer, and the amount of time that customers use a table. As a month progresses, a manager can spot a problem in sales or service by reviewing this report on a daily basis.

Food Cost Reports

The daily food cost sheet (see Chapter 20), prepared by hand, compares the cost of food against sales and determines a food cost percentage. This report can be computerized; the food cost report in Figure 21–5 (page 292), for example, reports the food cost per family, or menu, group. A daily food cost sheet and a computerized food cost report produce similar data. The report in Figure 21–5 breaks down the appetizer section of the menu into bar appetizers and dining room appetizers. It also lists the number of appetizers prepared in the kitchen and the number actually sold. If there is a difference, such as in dining room appetizers (one of the appetizer items was returned to the kitchen), then management is aware of a waste fac-

FAMILY GROUP FOOD COST REPORT MICROS 4700 VERSION 3.00 02-SEP-88 05:45 PM Page 1
ALL SYSTEM C: PRIMARY DATABASE

EXAMINE POSTED 02-SEP-91 05:29 PM CURRENT RESET 0002 02-SEP-91 02:32 AM
TO-DATE RESET 0001 01-SEP-91 02:35 AM

GRP	DESCRIPTION	CURRENT AND TO-DATE TOTALS						
		Qty Price	Qty Sold	Qty Rtrn	Total Prep Cost	Total Sales	Food Cost	Yield
1	Bar Apptzr	16 58	16 57	0 1	18.35 62.87	69.00 250.75	26.59 25.07	100.00 98.28
2	Dining Rm Apptzr	13 56	13 56	0 0	22.79 102.44	68.79 316.41	33.13 32.38	100.00 100.00
	SUBTOTAL	29 114	29 113	0 1	41.14 165.31	137.79 567.16	29.86 29.15	100.00 99.12
3	Soups	37 94	37 94	0 0	35.53 87.38	104.50 257.00	34.00 34.00	100.00 100.00
4	Salads	41 163	40 161	1 2	68.55 262.90	224.80 939.60	30.49 27.98	97.56 98.77

FIGURE 21–5 A Family Group Food Cost Report

tor. The total cost of preparation is given, along with total sales and a food cost percentage. The **percentage of yield** refers to the waste factor created by returned items. In the bar appetizer section, for example, one of the items was returned, which is reflected in the 98.28% yield versus the 100% yield for the other bar appetizer listed.

COMPUTERIZED PURCHASING RECORD KEEPING

The use of a computer to record purchasing transactions allows management greater access to price changes, updated inventories, and ordering needs. With computerized purchasing records, changes in food costs can be quickly noted and menu prices adjusted, if possible.

Purchasing Lists

The computer is valuable in tracking purchase orders and inventory as purchased goods travel through a restaurant (see Chapter 20). For example, a purchasing list can be supplied by a vendor to assist a purchasing agent in ordering food items. From this sheet, an order can be placed by telephone and recorded onto a computer (instead of written by hand). When the order is received, it can be checked in on the computer, which automatically lists the item as inventory.

Inventory Sheets

Using a computer, inventory can be updated and reviewed on a daily basis. Computerized inventories allow managers and chefs to accurately order food items as well as to keep track of food products still in inventory that will spoil if not used. A monthly inventory compares the amount of each item on hand at the beginning of a month against the amount on hand at the end of a month. Purchases made during the month are added so that the closing monthly inventory can be accurately valued. A computer can speed this process by quickly adjusting total values as the inventory changes.

COMPUTER-AIDED PRODUCTION REPORTS

Using a computer to produce recipe cards and cost cards allows management to easily adjust recipes, reducing or increasing the number of portions as needed. The immediate availability of adjusted recipes to a food production team increases quality control and keeps costs in line. A computer can also automatically change a recipe's cost card to reflect accurate costs.

Recipe Forms

The computer has become a valuable tool in the kitchen as well as in the

NUTRITION NOTE COMPUTERS IN INSTITUTIONAL FOOD SERVICE

Computers are especially useful in institutional food service, where meals must be planned to meet the nutritional needs of a particular group of customers as well as an organization's food cost requirements. Meals must be planned on a cyclical basis to repeat themselves over a period of time. These cycle menus require careful planning because some menu items, such as starches and vegetables, appear more frequently than others. Computers can use a menu file index to rotate menu items, providing variety to consumers. At the same time, computers can be used to plan in advance for purchasing and inventory needs to take advantage of volume purchase prices and seasonal food items.

Dietitians find computers extremely helpful in planning menus for people on special diets in health-related settings. From hospitals to geriatric-care centers, dietitians can use computers to quickly and easily plan menus for a wide variety of patient needs. For example, patient menus can be entered individually into a computer, which automatically calculates the amounts of different food items that are needed for a meal service. For these reasons, computers have been more widely incorporated into institutional food service than into any other area of food service production.

dining room and manager's office. For instance, when a server enters a customer order on a POS system, it prints the order on a remote printer in the kitchen. This system allows quick, accurate communication between the server and the kitchen, which results in better service to the customer. A computer can also act as a menu file storage system and generate recipe forms (much like the manually prepared recipe cards) on request. The recipe can be requested in whatever number of portions are needed for production. Quantities of ingredients are **exploded** (meaning increased) or reduced as necessary.

Figure 21–6 shows a recipe form for 50 portions of chicken cacciatore that was prepared on September 9, 1990, by Chef Ed Smith. The portion size is designated at 8 ounces of meat with 3 ounces of sauce. This recipe form includes the name of the product manufacturer indicated on the central inventory sheet, which lists all of the restaurant's food supplies and suppliers.

FIGURE 21–6 A Computer-Generated Recipe Form

PRODUCT: Chicken Cacciatore
DATE: 9/9/90
PREPARED BY: Ed Smith

Yield: 50 portions
PORTION SIZE: 8 ozs meat with 3 ozs sauce

PROD MFG ID	INGREDIENTS	AMOUNT	PROCEDURE
Sexton	Chicken	50 13-oz breasts	Clean and debone chicken. Dredge chicken in flour,
Sexton	Flour	24 ozs	salt, and pepper. Place
McCormick	Salt	1 oz	chicken fat down in hot oil.
McCormick	Black pepper	1 oz	Sauté 4 minutes on high heat
	Oil	1 lb	and on each side.
Fresh	Onions, cut 5/8"	3.5 lbs	Sauté fresh vegetables.
Fresh	Mushrooms, 1" button	3 lbs	Add tomatoes and spices.
Fresh	Green peppers, 1" strips	2 1/2 lbs	Bring to full boil. Add
Fresh	Garlic, fresh minced	3 ozs	sautéed chicken
Sexton	Tomatoes, crushed	14 lbs	and place in oven for 1 hour
McCormick	Oregano, crushed	2 tsp	30 minutes at 325°F.
Sexton	Salt	1/2 oz	
McCormick	Pepper	1/2 oz	
Sexton	Olives, green	2 ozs	
TOTAL WEIGHT:		66.125 LBS	
BREAST COOKED WT (60%)		24.375 LBS	

Cost Forms

The cost card that corresponds to each recipe card can also be generated by a computer, using information from a master file index. The information on the computer-generated cost form is the basis for the food cost data used on a daily food cost report. Figure 21–7 illustrates a cost form similar to one that would be generated by most food service software programs.

In Figure 21–7, the menu item is identified by the index number, name, and menu category. The selling price and the sales history (SLS, YTD) are listed for the current year to date. Each ingredient is listed and described, as is the quantity to be used. The cost of each ingredient is given by the unit of use (UU), such as one tablespoon or one ounce, and then the total cost as used in the recipe is calculated. Drawn butter, for example, is costed by unit of use at 5 cents per tablespoon, and, as used in the recipe, at 20 cents for four tablespoons. The total cost for one portion of the recipe comes to $4.38, which can be divided by the selling price of $15.95 to find the food cost percentage of 27 percent.

The cost information on this form changes according to the purchase price of each ingredient, which is posted and updated on a computer in

FIGURE 21–7 A Cost Form

Menu #101 Lobster Special Fish SP: 15.95 SLS, YTD: 23,006.35

Ingredient	Description	Quantity	Cost/uu	Cost	FC%
Lobster	1 lb lobster	1.00 each	4.00	4.00	
Lemon	Lemon wedge	2.00 each	.03	.06	
Butter	Drawn butter	4.00 tbsp	.05	.20	
Carrots	Baby carrots	3.00 oz	.03	.09	
Rolls	Dinner rolls	1.00 each	.03	.03	
		Total Cost:		4.38	
		Selling Price:		15.95	27%

the purchasing department. This integrated computer system results in accurate, up-to-the-minute information on cost forms, allowing management to see any major problems in food cost control.

CAREER PROFILE
ROBIN R. WEXLER: COMPUTER PROGRAMMER/INSTALLER, MICROS SYSTEMS, INC.

Robin Wexler began working in the food service industry when she was 15 years old. She held a variety of restaurant jobs throughout high school and college, gaining a wide and varied knowledge of the industry.

With a bachelor's degree in communications from Temple University, Ms. Wexler returned to the food service industry. The demands of the industry made her look toward areas of the business other than communications. She found an exciting new area in programming and installing food service computer systems for the MICROS Systems company.

Ms. Wexler states that her knowledge of restaurant and hospitality management is primary to being able to program. "The owners often do not know what the computer system needs to do," she says. "I tell them and write computer programs to fit the restaurants' needs in food and beverage operational controls."

Ms. Wexler often takes up to six months to install a computer system for a new or existing restaurant. The major functions of her job are to program a computer, train the staff to operate it, put it into working operation, and then be available to follow up on the operation of the system. Her next career step will be to go into direct customer sales of computer systems.

SUMMARY

The use of computers as a management tool in the food service industry is relatively new. The range of computer equipment available to the industry presents managers with many alternatives to recording, reporting, and monitoring their restaurant's business operations. Small pizza or sandwich shops rely on electronic cash registers to record transactions, for example. Fast-food operations assist servers with specially designed display panels. Family style operations use computerized cash registers for basic record keeping. Restaurants that want a complete system of ordering panels, remote printers, and management software programs take advantage of modular computer systems.

From guest checks to payroll registers, managers can now rely on computer-based information systems to quickly and accurately print hard copy and generate service, purchasing, and production reports. It is important to remember, however, that a computer is just a piece of equipment, and it will provide only the information that it is programmed to process. The quality of the software program used in a POS system controls the quality of the report forms it produces. If inaccurate data is entered into a computer, the resulting forms will be wrong. Management must constantly be aware of these factors when handling computer systems.

QUESTIONS

1. In what ways do computers ease management's responsibilities?
2. A computerized cash register can record what type of information?
3. Define the term *POS system*. What is the major feature of this computer system?
4. Outline the modular computer system discussed in this chapter. What is the flow of information from one workstation to another?
5. How is manual organizational record keeping different from computerized organizational record keeping?
6. What reports receive information from guest checks?
7. What does the term *table turn* mean?
8. Why is it important to present total daily sales as well as total to-date sales on an employee sales profile report?
9. What does the term *exploded* mean as it relates to a computer-generated recipe form?
10. By using a computer to calculate cost forms, management can be constantly updated on what important information?

ACTIVITIES

Activity One Compare the computerized operational reports presented in this chapter with their manual counterparts in Chapter 20. Although their formats are different, how are the types of information they provide similar? Explain in what ways the computer-generated records are more effective for management use.

Activity Two Using pictures from magazines or brochures, make a chart that identifies the equipment discussed in this chapter.

Activity Three Visit some food service operations in your town or city. Identify the different types of computer systems in each restaurant. Report to your class on the variety of systems being used in your community and the types of food service operations that use them. Obtain, if possible, samples of different report forms.

Activity Four With the help of your teacher, investigate the food service operation in your school. Is it computerized? What types of equipment and programs are being used to record sales? Are purchasing and inventory computerized?

22

LABOR RELATIONS

VOCABULARY

job description
job specification
structured interview
semistructured interview
unstructured interview
job orientation
delegation
compensations
salary
wages
Fair Labor Standards Act of 1938
Social Security Act of 1935
Worker's Compensation Act
Retirement Income Security Act of 1974
Federal Wage Garnishment Act
Polygraph Protection Act
performance evaluation

OBJECTIVES

After studying this chapter you should be able to do the following:

- Identify the six responsibilities of management and understand how they apply to promoting good labor relations in the workplace
- Identify recruitment sources for new employees
- Identify the special groups of people that make up the labor pool for hospitality and food service positions
- Develop a questionnaire that can be used during an interview
- Explain the labor laws that relate to employers in the food service industry

Labor relations involve the staffing and managing of human resources to accomplish certain goals. Employers or managers must follow and enforce the fair labor policies required by law without discriminating against employees because of race, religion, national origin, sex, or limits to physical or mental abilities that do not affect work performance. The effective management of human resources is the most important factor contributing to the success of food service organizations.

Employers and managers must be well trained in human relations as well as technical skills to be able to carry out all management responsibilities. These duties make up the cycle of management responsibilities, which includes planning, organizing, staffing, directing, controlling, and evaluating the human and technical resources available. Good labor relations between employers and employees and fair labor practices reduce absenteeism, increase productivity, and provide an overall positive working environment.

PLANNING AND ORGANIZING

Planning and organizing are the first two duties in the cycle of management responsibilities (see Figure 22–1). A good plan makes a task easy to understand and complete. Planning involves using one's head to think about what needs to be accomplished. Organizing is putting a plan in a logical and systematic order. It helps to start the planning process by writing thoughts down. When planning and organizing, it is important to ask the questions listed in Table 22–1 (page 300).

Planning and organizing in a logical manner makes it easy for employees to follow directions. In addition, good planning and organizing can help divide the workload evenly, which leads to positive communications between employees and between employees and management. Employ-

FIGURE 22–1 The Cycle of Management Responsibilities

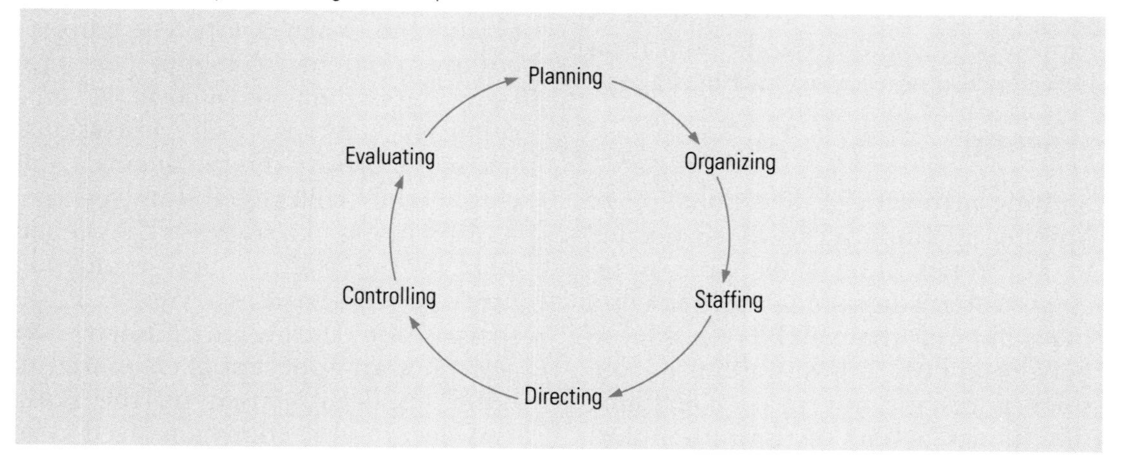

PLANNING QUESTIONS	ORGANIZING QUESTIONS
What do I want to accomplish?	How many labor hours are needed?
How will I accomplish it?	How many employees do I need?
When will I accomplish it?	What type of equipment do I need?
Why does it need to be accomplished?	What supplies do I need?
Who will accomplish it?	What is the cost of supplies?
	What is the cost of labor?

TABLE 22–1 Planning and Organizing Questions

ees feel more secure and confident when they know what they have to do and how to do it. Systematic planning and organizing increases productivity, reduces absenteeism, and helps to orient all the available resources toward common objectives. In a food service organization, this concept of teamwork is important. If employees do not work as a group, the work environment will be negative, and this negative attitude will be passed on to customers, who will eventually take their business elsewhere. Figure 22–2 illustrates an organizational chart of restaurant employees working together for the same goal. The lines represent the lines of communication between the employees and managers.

STAFFING

Staffing involves the determination of how many employees will be needed to get a job done. This process includes forecasting the labor hours and the skill levels that will be necessary to perform a task. It is important to consider a restaurant's peak hours and slow hours before recruiting employees. Peak hours are when an operation is very busy and more staff is needed, such as during breakfast, lunch, dinner, and the weekend. Less staff is needed for slow hours, such as the midmorning, the late afternoon, after dinner, and, in some parts of the country, the summer months. For example, a university food service facility does less business during the summer, since many students have left the campus. During these months, staffing needs are minimal.

Recruitment

Recruiting new employees can be done from within a food service organization, as when competent employees are promoted to higher positions, or by external means, such as personal contacts, newspapers, professional journals, private employment agencies, schools, colleges, and universities.

Recruiting employees is the second step in the staffing and evaluation processes outlined in Figure 22–3 (page 302). During recruitment and selection, an applicant is compared to other applicants based on the skills, knowledge, and education needed to perform a job as outlined in a job description and a job specification. A **job description** and a **job specification** give detailed information about a job as well as the skills and qualifications needed to do it, allowing both employers and applicants to fully understand the position.

The application form that a food service organization uses to recruit and screen applicants should be phrased clearly and without discriminating questions. It should also, however, give an employer the necessary information to help make a final selection. Requesting a résumé in addition

ILLUSTRATION 22-1 Requesting a résumé in addition to an application gives an employer additional information about an applicant.

FIGURE 22–2 Organizational Chart of Restaurant Employees Working for the Same Goal

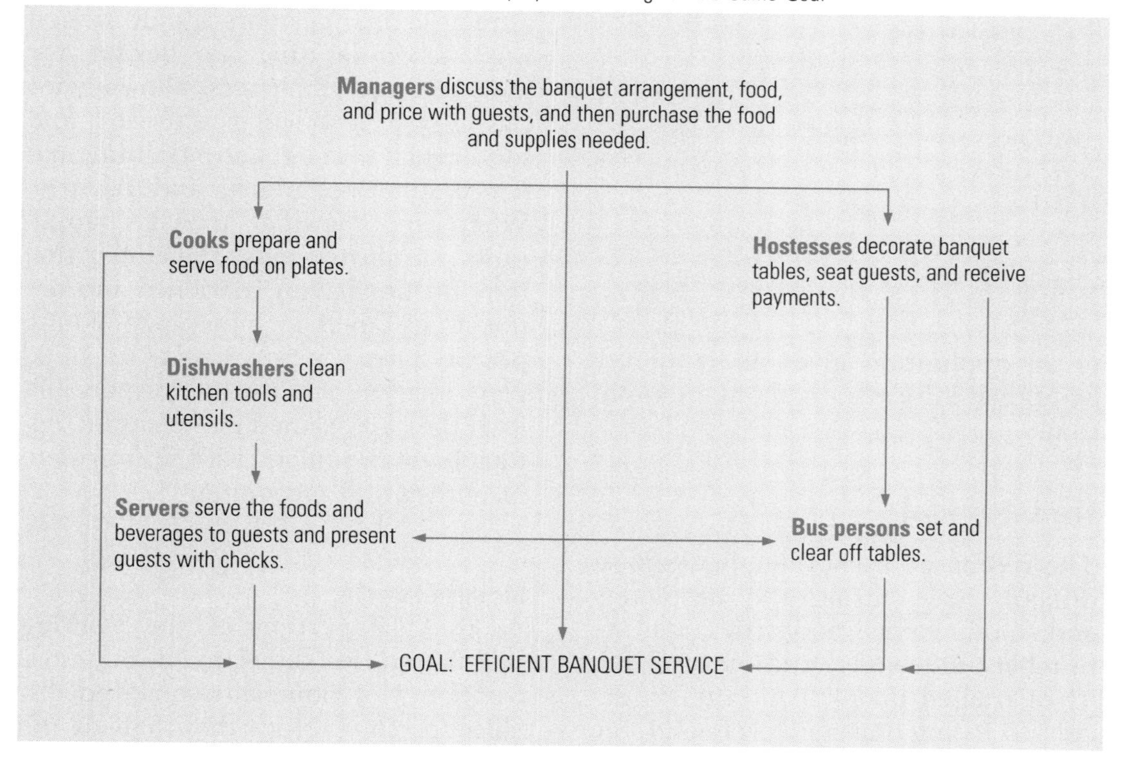

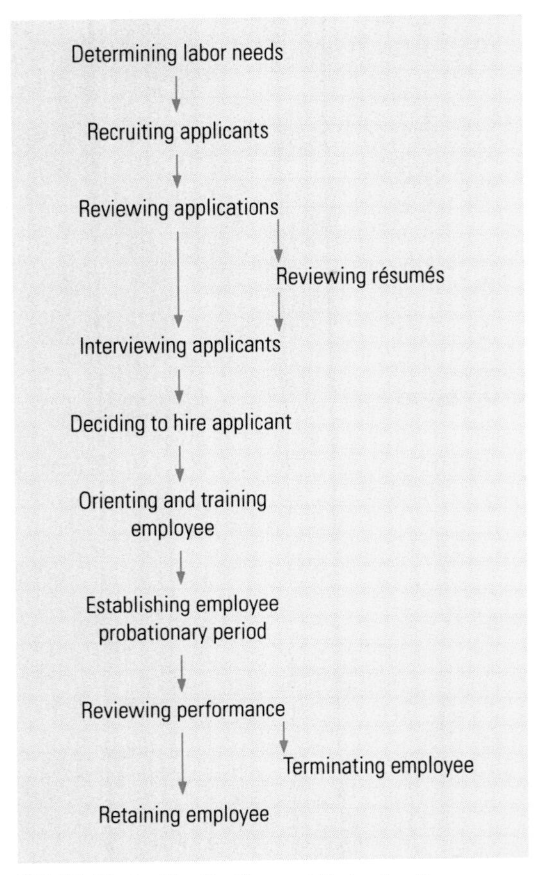

FIGURE 22–3 The Staffing and Evaluation Processes

to an application gives an employer additional information about an applicant.

The Labor Pool

The recent labor shortage is the most difficult issue facing the hospitality and food service industries, as the number of teenagers and the availability of trained people continues to decrease. Rising income levels and the growing number of men and women in the work force have contributed to the fast growth in food service employment, including many new positions for chefs, cooks, bartenders, and servers. In the 1970s, younger employees—teenagers—were the primary workers in the hospitality and food service industries. Since the mid 1980s, the number of teenagers has declined and the demand for older employees has increased steadily.

Teenagers

In 1985 teenage workers constituted 12 percent of the food service labor force. Since that time, the percentage of teenage workers has declined. To attract teens, managers should design work schedules that are flexible for their teenage employees, allowing for classes, extracurricular activities, and socializing. Teens require training, patience, time to adjust, and fairness. When interviewing teenage applicants, employers must remember that they do not have experience and are not accustomed to interviews. Young people have a great capacity to think, plan, analyze, and assume responsibility if they are trained in a pleasant, equitable environment with appropriate incentives and recognition.

Disabled Persons

Around 19 percent of the working population in the United States is composed of noninstitutionalized disabled persons. More than half of the disabled who could work, however, are without jobs because of prejudice and numerous educational and training barriers.

Recent legislation has attempted to protect the disabled from discriminating labor practices. Section 190 of the Tax Reform Act of 1986 provides up to $35,000 in tax credits annually for employers who make property improvements to accommodate disabled persons. In addition, many state agencies provide funds to compensate employers for part of the time spent training disabled employees in on-the-job-training (OJT) programs. State funds are also available for supplies and special equipment needed by employers who hire the disabled. Many hotel and food service employers who have hired and educated disabled employees through classes and OJT programs have found them to be productive and loyal.

Mature Workers

To respond to the labor shortage, many employers in the hospitality industry are recruiting mature workers. Mature workers (those over 65 years of age) are increasingly available and eager to work. Mature persons are excellent candidates for employment in the hospitality industry because they have a lot of experience and tend to be productive. In addition, the Job Tax Credit provides employers a 50

ILLUSTRATION 22-2 Many hotel and food service employers who have hired and educated disabled employees through classes and OJT programs have found them to be productive and loyal.

ILLUSTRATION 22-3 Mature workers (those over 65 years of age) are increasingly available and eager to work.

percent tax credit on a mature worker's first $6,000 in wages for one year.

The Job Application

A job application form is an important tool used by employers to screen job applicants. It identifies the position that an applicant is interested in; provides an employer with the basic information needed to conduct a job interview, such as personal data, previous work experiences, education, special skills, and reference contacts; and also serves as part of the organization's personnel records if an applicant is hired.

Inquiries about age, sex, race, religion, or national origin violate an applicant's civil rights and cannot be included on an application form. Once an application is filled out by an applicant, the employer or interviewer should read it thoroughly and check for the following important details: neatness, spelling errors, education, references, completeness, and signature.

The Résumé

A résumé is a summary of a person's experiences and qualifications for a job. It should give an employer information that interests her or him enough to call the applicant for an interview. A résumé should take less than two minutes to read and should be no more than two pages. Long résumés are time consuming and are usually tossed aside by an employer. A good résumé has as few words as possible and at the same time gives all the appropriate information. A résumé is intended to highlight the applicant's qualifications and to make an employer want to learn more about the applicant's strong points through an interview. Although there are many different formats that can be followed when writing a résumé, it should include the following information: name, address, career objective, educational background, employment record, personal data, and references.

The Interview

An interview is the method most widely used by employers to assess an applicant's qualifications. A person conducting an interview uses an applicant's job application and résumé as guides for formulating the questions asked during the interview. The main reason for an interview is to get information on an applicant's background, attitudes, feelings, and personality traits that may influence the job selection process.

In a personal interview, an interviewer has the opportunity to become acquainted with an applicant and to observe personal characteristics that are not evident in an application or résumé. During an interview, an applicant should be treated with courtesy and respect.

There are three different types of interviews. In a **structured interview,**

ILLUSTRATION 22-4 An interview is the method most widely used by employers to assess an applicant's qualifications.

an interviewer asks all applicants the same questions, prepared ahead of time, and records each applicant's responses. This type of interview is limited to necessary information and minimizes possible bias or prejudice by an interviewer. In a **semistructured interview,** only some of the questions asked by an interviewer are prepared ahead of time. This type of interview is more flexible for an interviewer because it allows her or him to ask additional questions that she or he may find necessary for the selection process. An **unstructured interview** gives even more freedom in terms of questioning. An interviewer may ask broad questions, such as "tell me about yourself." This type of interview is useful when personal characteristics are important parts of a job description. Table 22–2 (page 306) lists topics that, by law, must be avoided by an interviewer during any type of interview.

The Decision to Hire

A decision to accept or reject an applicant must be made carefully. Once a final decision is made, a new employee may be required to have a physical exam. The reasons for a physical

Name
Questions regarding an applicant's ancestry, national origin, descent, or titles, such as Miss, Mrs. or Mr., are forbidden.

Address
An interviewer cannot inquire into a foreign address that might indicate an applicant's national origin.

Age
The Age Discrimination in Employment Act of 1967 forbids an employer to discriminate against applicants between the ages of forty and seventy. Asking the applicant's age, date of birth, or proof of birth is not allowed by law.

Birthplace
Questions about where an applicant, spouse, or other family members were born are forbidden.

Religion
Questions about an applicant's religious preference, observation of religious holidays, or membership in a church or parish is forbidden. An employer may not tell an applicant that only certain religious groups can observe religious holidays.

Sex
Asking about the sex of an applicant or questions that may indicate that the employer is inquiring about the sex of the applicant is forbidden. Sex is not a factor in job performance. Height and weight requirements, however, should be stated in a job description.

Race or Color
The race or color of the applicant's skin may not be questioned directly or indirectly by an employer because they do not relate to job qualifications.

Handicaps
The Rehabilitation Act of 1973 forbids employers from asking applicants questions regarding a handicap or the severity of a handicap. Employers must make reasonable changes to accommodate the physical and mental limitations of handicapped individuals.

Marital Status
An interviewer may not inquire whether an applicant is married, single, or divorced. Asking about the number of children, child care arrangements, or intent of having children is also against the law.

Citizenship
An interviewer may not ask the date an applicant became a U.S. citizen, of what country she or he is a citizen, if her or his parents or relatives are naturalized, or if an applicant has naturalization papers.

Education
Inquiries as to how an applicant learned to speak a foreign language, or other questions that might indicate that an employer seeks information on nationality, religion, or race, are forbidden.

Convictions or Arrests
An interviewer cannot require an applicant to provide arrest or court conviction records when they are not related to the functions and responsibilities of a job.

Credit Rating
Any questions regarding an applicant's credit rating, charge account, and ownership of car are forbidden.

Military Record
An interviewer may not ask about the type of discharge that was given to an applicant who was in the military.

TABLE 22–2 Topics to Be Avoided by an Interviewer

exam are to determine if the new employee can meet the physical demands of the job, to provide a health record that will protect the organization against claims for injuries that were caused prior to the employee's hiring, and to identify any disease that the employee may carry that could be transmitted through food handling or through contact with other employees.

Job Orientation

Job orientation is the most important phase in staffing, and it should be designed to interest a new employee, familiarize her or him with the business, and gain her or his support for organizational goals and objectives. A thorough orientation program includes the history and goals of the organization, policies, procedures, benefits, employee expectations, introductions to other employees, and a tour of the facility.

DIRECTING

Once a food service business's planning, organizing, and staffing needs have been met, the next step on the management cycle is directing. Directing is, simply, the delegation of responsibility and authority.

Delegation

Delegation is the act of giving another person the responsibility for a job and the permission to perform it. Before delegating a task to an employee, however, a manager should first discuss the job with the employee and then delegate a supervisor to train or instruct the employee. The supervisor explains the delegated job and the preferred method of doing it.

Job Instruction and Training

It is a supervisor's responsibility to make sure that an employee understands a job and can demonstrate his or her abilities. A supervisor can make sure that an employee has mastered a task by asking the employee questions while instructing her or him on how to perform the task.

In food service institutions, OJT programs are essential to the success of an organization. The objectives of food service OJT programs are to teach employees to serve high quality food at reasonable prices as well as to avoid accidents and damage to property and equipment. Before a manager, supervisor, or trainer can instruct an employee on how to perform a particular job, he or she must have a job analysis that breaks down the job into one systematic operation.

CONTROLLING

A food service manager is responsible for controlling or exerting influence over all the costs necessary to operate a business and make a profit. The following sections deal with controlling labor costs and increasing productivity through scheduling, communication skills, compensation, labor laws, and issues of sexual harassment. Other areas of cost control, such as food purchasing, food storage, and recipe standardization, are discussed in Chapters 12 and 13.

Scheduling

Scheduling is an operational tool used to control human resources effectively. There are two basic methods of scheduling:

1. Vertical scheduling—The work is divided into separate tasks, and each employee is responsible for a different task.
2. Horizontal or assembly line scheduling—The work is divided into steps, and each employee is responsible for a different step.

The horizontal method of scheduling is more effective in large institutions, such as hospitals and large school district food service operations. Employees working in a horizontal plan are more flexible because they can be trained to do a variety of different tasks.

Before attempting to write a schedule, a manager must first determine how many employees are needed at each meal service, at what hours. It may be necessary to consider hiring part-time employees for peak hours. Experts recommend that an average meal should take between 14 and 17 minutes of labor time from the beginning of food preparation through service and cleanup. Larger amounts of labor time per meal should include performing more tasks, such as serving multiple courses or cleaning up difficult messes.

Communication

Communication is the most important skill that managers must have in order to lead their employees. Many food service managers today fail to gain employee cooperation because they are unable to communicate effectively with their employees. They lack enthusiasm, have negative attitudes toward their employees, are poor listeners, and have little tact when asking employees to do a job.

In order to be successful, a food service manager must be interested in the job she or he is doing and must have the ability to work in harmony with people. A good manager has an honest and straightforward manner, objectivity, and maturity. A manager needs to sell her or his idea of work to employees in order to gain their cooperation by

- asking employees to play an active role in suggesting how an idea will be carried out.
- showing employees that he or she can make mistakes just like other employees, and being willing to admit those mistakes.
- showing appreciation for a job well done and praising employees whenever they deserve it.
- avoiding arguments with employees in front of fellow employees, causing embarrassing scenes.
- acknowledging and rewarding employees when they perform above and beyond their job responsibilities.

Compensation

Employee salaries, wages, and benefits are the **compensations** received by employees for work done. Labor costs, in the form of compensation, make up

MATH SCREEN

A hospital food service facility serves 250 meals for breakfast, 250 meals for lunch, and 250 meals for dinner each day. These meals are served at the following hours:

Breakfast	7:00 a.m.–7:30 a.m.
Lunch	11:30 a.m.–12:00 p.m.
Dinner	4:30 p.m.–5:00 p.m.

In an effort to keep the labor time per meal below the average, management decides to lower labor times from 15 minutes of labor per meal to 12 minutes of labor per meal.

Use the following information for calculations:

Labor cost = $6 per hour

Full-time work = 8 hours per day
Part-time work = Less than 8 hours
Overtime work = More than 8 hours

The following equations show how to calculate total labor hours:

250 meals × 3 meals per day = 750 meals per day

750 meals × 12 minutes per meal = 9,000 minutes

9,000 minutes ÷ 60 minutes = 150 hours

The following chart lists the hospital's labor needs and demonstrates how to calculate total hours for each position and for the entire staff:

Staff	Number of Labor Hours		Number of Employees		Total Hours
Overtime kitchen supervisor	10	×	1	=	10
Full-time floor supervisors	8	×	2	=	16
Full-time cooks	8	×	2	=	16
Part-time cooks	4	×	2	=	8
Full-time kitchen helpers	4	×	2	=	8
Part-time kitchen helpers	4	×	2	=	8
Overtime dishwashers	10	×	2	=	20
Full-time service people	8	×	4	=	32
Part-time service people	4	×	6	=	24
					142

To calculate labor costs, add the full-time and part-time labor hours and multiply by the regular hourly rate. Then add overtime labor hours and multiply by the time and a half hourly rate:

136 labor hours at regular rate
136 × $6 = $816
6 labor hours at overtime rate
6 × $9 = 54
$870

To calculate the labor cost per meal, divide the total cost of labor by the number of meals served:

$870.00 ÷ 750 = $1.16

Now test your math skills:

1. How many full-time labor hours are required to prepare and serve 150 meals at 16 minutes per meal?
2. How many employees are needed to prepare the meals in problem 1 if all the employees worked full-time?

a substantial part of the total costs of an operation. **Salary** refers to the amount of money received by employees in management and professional positions. **Wages** refer to money earned by the hour. Employees working for hourly wages are covered by the Fair Labor Standards Act, also called the Minimum Wage or Wage-Hour Law. Benefits are other forms of compensation received by employees in addition to salary or wages. From the 1930s on, federal law has made many significant additions to employee compensation, including unemployment compensation, nondiscrimination compensation, retirement compensation, baseline (minimum wage) compensation, and garnishment laws. These additional forms of compensation have increased benefit costs to employers.

Labor Laws

The **Fair Labor Standards Act of 1938** established the minimum age and overtime requirements. Food service employees were not covered by the act until the mid 1960s. The Fair Labor Standard Act requires employers to

- pay employees time and a half for all hours worked in excess of 40 hours per week. Employees on salary compensation, such as executives, administrators, and most professional employees, are usually exempt from this law, depending on their job responsibilities and salary level.
- hire employees over the age of 18 for hazardous jobs. It also forbids the hiring of employees under the age of 16.

NUTRITION NOTE EMPLOYEE HEALTH

Good nutrition and physical fitness are essential components of an unstressed and pleasant working environment. Research has shown that employees who eat a balanced diet, including foods from the meat, dairy, grain, and vegetable and fruit groups, and participate regularly in an exercise program have a more positive attitude toward their job than employees who do not exercise or eat well. Today, more and more employers provide access to free or inexpensive healthful meals as well as to the use of exercising facilities or programs as part of their benefit packages. The cost of these benefits is small in comparison to the returns gained from employee satisfaction. Employees who are given such benefits come to work with high morale, make friends in the workplace, and are more cooperative and loyal to management.

The **Social Security Act of 1935** was established to protect employees against the loss of earnings after retirement, unemployment, disability, or the death of a supporting spouse. The program is supported by a tax withheld from employees and an amount paid by employers.

Unemployment compensation laws vary from state to state. These laws state that working employees covered by the Social Security Act who have been involuntarily laid off and are unemployed for a period up to six months must be provided weekly compensation by the state. The amount paid depends on previous earnings and length of employment before termination.

The **Worker's Compensation Act** provides compensation to workers injured on the job. This law requires employers to

- provide income and medical benefits to employees injured in work-related accidents.
- provide safety and accident pre-

ILLUSTRATION 22-5 Unemployment compensation laws state that working employees covered by the Social Security Act who have been involuntarily laid off and are unemployed for a period up to six months must be provided weekly compensation by the state.

vention training programs and encourage safety in the workplace at all times.

The **Retirement Income Security Act of 1974** protects employee pension plans by restricting employers' right to operate and control them. This act is sometimes called the Pension Reform Law.

The **Federal Wage Garnishment Act** protects an employee's income by limiting the amount that may be garnished. Garnishment is a legal action requiring an employer to withhold the earnings of an individual employee for payment of a debt.

The **Polygraph Protection Act** states that restaurant operators and other employers cannot use lie detectors to screen job applicants. This law also prohibits an employer from asking an employee to take a polygraph test unless the employer is investigating a case of theft in which the employee had access to the property that was stolen.

Sexual Harassment

Sexual harassment in the workplace is a violation of Title VII of the Civil Rights Act of 1964. Although sexual harassment is a relatively new legal issue, it has been treated with seriousness and concern by courts, government agencies, businesses and industries. Employers who do not take appropriate action when they know that a supervisor or employee has made sexual demands on another employee are in violation of the Civil Rights Act and subject to a fine in the form of employee compensation.

Sexual harassment causes employee dissatisfaction, decreases productivity, and increases absenteeism in the workplace. It often results in listlessness, headaches, stomach problems, decreased concentration, loss of ambition, and depression.

EVALUATING

A **performance evaluation** is a written record that serves many purposes, which include providing employees with feedback on the acceptability of their work, encouraging and praising employees for a job well done, and documenting information to help employers make future decisions about job assignments and promotions. The criteria used to measure an employee's performance should be relevant to the job, understandable, and easy to use.

There are several methods of evaluating employee performance. The most widely used evaluation procedure is the rating method. Using this method, an employee is rated from one (poor) to five (superior) on several criteria, including attendance, initiative, the quality and quantity of her or his work, and other personal characteristics that are related to job performance. An employer must discuss a performance evaluation with the employee being evaluated, and the evaluation must be signed by both parties.

CAREER PROFILE
POSITION: PERSONNEL MANAGER

Qualifications:

- Good verbal, math, and writing skills, and the ability to communicate with employees effectively
- Two years experience in personnel management, payroll processing, and employee recruitment and hiring
- An equivalent combination of training, education, or experience that meets the above qualifications

Responsibilities:

- Develops hiring, training, and evaluation procedures for new employees.
- Recruits, interviews, and hires new employees
- Prepares employee work schedules
- Assigns and evaluates the work of the assistant clerk typist
- Prepares paperwork reflecting any personnel or payroll changes
- Maintains sick leave and vacation balances for employees
- Handles employee complaints and problems
- Conducts regular employee meetings dealing with personnel issues

SUMMARY

To establish effective labor relations, an employer must master certain duties. These key management responsibilities include planning, organizing, staffing, directing, controlling, and evaluating.

After determining their labor needs through the planning and organizing processes, employers must recruit and hire appropriate staff members. The job application and the résumé are important tools available to help employers select new employees. When these tools are used effectively during the interview process, they can help interviewers make final selections for the hiring of new employees. Employers who understand labor laws and have good communication skills are able to direct, control, and evaluate their employees with fairness, objectivity, honesty, and appreciation. These management skills can increase employee satisfaction and productivity.

QUESTIONS

1. What are the first two responsibilities in the cycle of management responsibilities?
2. What are five sources from which new employees may be recruited?
3. What are two tax incentives for employers who hire or train disabled employees?
4. What two important tools are used by employers to screen job applicants?

5. In what type of interview does an interviewer ask all applicants the same questions, prepared ahead of time, and record the applicants' responses?
6. What type of interview gives an employer freedom to ask questions that are broad in nature and that are not the same for all applicants?
7. What types of questions are in violation of the Civil Rights Act if they are asked by an employer on an application or during an interview?
8. What is the act of giving a person the responsibility and permission to do a job?
9. What operational tool is used to control human resources effectively?
10. What is the most important skill that managers must have to gain cooperation from employees?
11. What term describes the salary, wages, and benefits given to employees for their work?
12. Which federal act was responsible for establishing minimum wage and overtime requirements?
13. Which federal act protects employees from losing their earnings after retirement, unemployment, disability, or the death of a supporting spouse?

ACTIVITIES

Activity One Plan a banquet dinner for 25 people by using the planning and organizing questions in Table 22–1 (page 300) to determine what resources are needed.

Activity Two Write a set of questions that can be used to interview an applicant for the management position described in this chapter's Career Profile box. Discuss the questions in class, using Table 22–2 (page 306) as a guideline.

23

MARKETING AND MERCHANDISING

VOCABULARY

marketing
marketing cycle
promotions
merchandising
marketing mix
media
public relations

OBJECTIVES

After studying this chapter you should be able to do the following:

- Differentiate between the concepts of marketing and merchandising
- Chart the path of the marketing cycle
- Compare and contrast the food service marketing mix with the traditional marketing mix
- Create a food service marketing campaign

Accounting for over $227.3 billion in annual sales and 8 million jobs, the food service industry is a growing part of the American economy. Recent changes in Americans' work and leisure life-styles have affected the way in which food service businesses are used by the general public. The rise in single-parent and double-income families has resulted in an increasing demand for pre-prepared foods and meals cooked outside of the home. The American public's changing eating patterns require major adjustments on the part of the food service industry, which must meet the public's needs and demands. The challenge to create new products and services as well as to advertise their availability is called marketing and merchandising.

MARKETING AND MERCHANDISING

Identifying customer needs and developing products and services to satisfy those needs are entrepreneurial activities (which are discussed in detail in Chapter 19). Marketing and merchandising are business activities that can help an entrepreneur to successfully identify and develop products and services.

Marketing

Marketing involves business activities that direct the flow of goods and services from producers to consumers; it is any function that affects sales. Marketing may be defined as human activity aimed at satisfying consumer needs and wants through the exchange processes. In terms of the food service industry, these "exchange processes" are money for food products and the services needed to provide them. Because different types of services are of primary importance in the food service industry, the challenge of marketing is to develop not only food products but the expectations, entertainments, and services that surround them.

The Marketing Cycle

The **marketing cycle** is a series of steps that must be followed to successfully identify a need, develop a product, generate customer interest, and, lastly, evaluate financial success and customer satisfaction. Figure 23–1 shows the marketing cycle as it relates to the food service industry, and the following sections describe each step in this process.

Identifying a Need

A customer need is usually realized when management recognizes that they are not providing a service or a product that has either been requested by a customer or is being offered by other food service operations

FIGURE 23–1 The Marketing Cycle

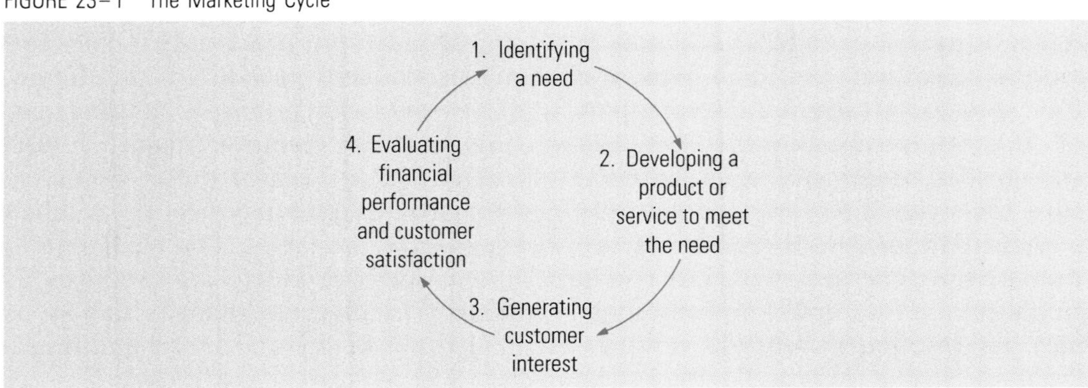

and is meeting some measure of success. For example, a customer request to celebrate Father's Day in a restaurant setting led a restaurant's managers to see that they could satisfy a widespread need by offering such a service.

Developing a Product or Service

After a need is identified, a product or service must be developed to meet that need. Continuing the preceding example, the restaurant's managers responded to their customers' need for a Father's Day meal by analyzing their menu for items that would be appropriate for a family celebration. They identified a dinner selection that would appeal to their customers. A special price of $9.45 was assigned to a three-course dinner consisting of salad, steak, and strawberry pie to be served from noon until closing time on Father's Day.

Generating Customer Interest

Creating customer interest can be done through in-house promotions or outside advertising. **Promotions** are activities developed and conducted to increase the growth of a business or the desire for a service. Promotions for the special Father's Day meal, for example, were developed around the customer need to celebrate Father's Day in a restaurant setting. A promotional card was attached to each menu or displayed in a menu stand on each table for two weeks prior to Father's Day.

ILLUSTRATION 23-1 One restaurant's managers responded to their customers' need for a Father's Day meal by assigning a special price to a three-course dinner consisting of salad, steak, and strawberry pie.

Evaluating Success

The success of a new product or service is evaluated by the level of customer satisfaction and by financial performance. To determine financial success, the total sales for a day can be balanced against costs to indicate the amount of profit generated by an item or service. Customer comment cards combined with server feedback and management's observations can determine customer satisfaction. For example, the Father's Day meal was deemed a success because it generated a solid profit and servers reported that customers seemed satisfied with the

NUTRITION NOTE PROMOTING SEASONAL FRUITS AND VEGETABLES

Promotional activities in food service operations often center around a seasonally available fruit or vegetable. A variety of menu items are offered that use a popular food item as the central ingredient. For example, strawberries, which are rich in Vitamin C and fiber, are a popular promotion in early summer, when they are available in large quantities at reasonable prices. Using fresh fruits and vegetables at the peak of their nutritional value can add fresh, natural, and nutritious items to menus.

Fresh fruit can be used to create attractive and colorful desserts. Examples from a menu follow.

meal. If both financial success and customer satisfaction are noted in a restaurant's daily sales history, then future attempts at a particular promotion will be more effective.

Merchandising

Merchandising is sales activity that generates customer interest. These activities include a wide range of promotional items or events as well as advertising in newspapers, magazines, radio, and television.

At the end of the promotional time period, the success of a merchandising program is evaluated by totaling the sales of the items featured in the program (or promotion) and comparing this figure to the overall sales figure. If there is a significant increase in overall sales and a healthy profit is realized, then the program is considered successful. If, on the other hand, the overall sales do not increase, then management should reconsider repeating the program.

Because most merchandising activities require significant printing and advertising costs, managers should research marketing concepts thoroughly before putting them into action. Accurately evaluating a product or service's need and market are necessary steps in developing a successful product or service and merchandising it well.

ILLUSTRATION 23-2 Merchandising, such as this restaurant advertisement, is sales activity that generates customer interest.

THE FOOD SERVICE MARKETING MIX

In developing new products and services, management considers four marketing elements that are concerned with customers' needs. These elements are called the **marketing mix,** and they include product (or service), place, promotion, and price. These four elements address all aspects of customers' needs: the product or service itself, the public setting in which it is offered, the way in which the general public is informed about the product or service, and the price it is given to ensure that customers feel it has a fair value.

Because the food service industry creates products that are different from other types of products, the traditional marketing mix is changed to reflect these differences. Because a food service product cannot be presented without service, the first marketing mix element is changed to product-service. Place means more than location in food service, and, therefore, has become presentation, which includes atmosphere and service. Finally, promotion and price are combined into communications, reflecting the integral part that communication plays in all aspects of customer relations. Figure 23–2 compares traditional and food service marketing mixes. Food service managers should base their decision making on the elements of the food service marketing mix to increase the success of their marketing efforts.

A SAMPLE MARKETING CAMPAIGN

Developing a marketing campaign (see Figure 23–1 on page 316) includes identifying a customer need, creating a product or service, generating customer interest in buying the product or service, and, finally, evaluating success. This section of the chapter follows the path of a marketing campaign through these four steps.

Identifying a Need

Throughout this book we have discussed a growing customer concern for the availability of healthful foods

FIGURE 23–2 Traditional and Food Service Marketing Mixes

TRADITIONAL MARKETING MIX	FOOD SERVICE MARKETING MIX
Product	Product-service (The combination of products and services in a food service operation)
Place	Presentation (The exterior and interior decor, location, personnel, and menu pricing structure)
Promotion Price	Communication (The menu, server communication, advertising, promotion, and customers' comments and observations)

in all types of food service operations. The variety of dietary needs ranges from low-calorie to low-sodium meals, depending on the health concerns of the customer.

The American Heart Association, restaurant associations, and hospitals have noted that research data (such as those presented in the graph in this chapter's Math Screen) confirm that

MATH SCREEN

The first step in the marketing cycle (see Figure 23–1 on page 316) is to identify what a consumer wants. Research data such as those presented in the following marketing research graph provide the information necessary for food service marketers to formulate ideas and concepts that may result in new products and services.

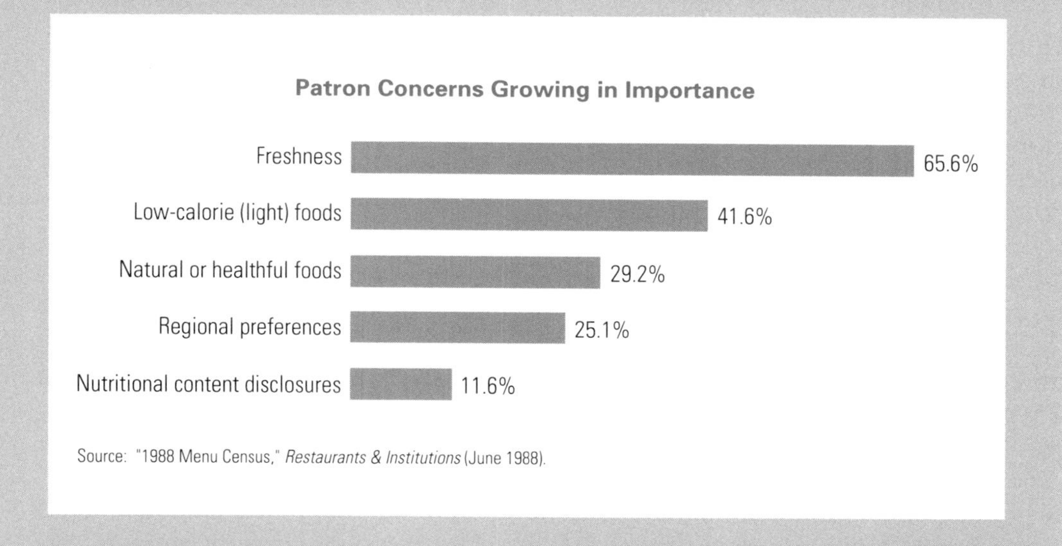

Source: "1988 Menu Census," *Restaurants & Institutions* (June 1988).

Use the graph to answer these questions:

1. Of the patrons surveyed, what percentage show a concern for being served low-calorie (light) foods?
2. Did the survey allow each patron to express only one concern, or were the patrons allowed to state more than one concern? How do you know?
3. If 800 restaurant patrons were surveyed, how many of them selected freshness as a growing concern?
4. Again assuming 800 respondents, which category was selected by 234 people as a growing concern?

being able to eat healthy foods outside of the home has been a growing need of many customers. Using similar research, customer feedback, and input from the medical community, the food service industry has also recognized this need and has given it a general marketing identification: Health & Wellness. The first step in this sample marketing campaign has thus been completed: a customer need has been identified.

Developing a Product or Service

In the second phase of the marketing campaign, different organizations in the food service industry had to decide whether the Health & Wellness need could be met by developing a product or a service. As Figure 23–2 (page 320) shows, the food service marketing mix combines these elements into one element, the product-service.

Restaurant associations, hospitals, and the American Heart Association have developed a joint venture called the Dining With Heart program. The object of the program is to adjust current recipes or to develop new ones for restaurants that want to respond to one or more customer health needs. The program is based in the nutrition and dietary departments of participating hospitals. Registered dietitians meet with restaurant chefs to determine how to change or create low-calorie, low-sodium recipes.

The final aspect of this product-service is to highlight healthful meals by including them on restaurant menus with a Dining With Heart symbol. In addition, servers are given the nutritional breakdown of these items so that they can answer customer questions.

Mixed Fruit Salad 1.75
Ripe fruits in season, chilled in their own juice.
Chopped Chicken Livers 2.95
With lettuce, Spanish onion, tomato slices.
Sweet and Sour Meatballs 3.95
Petite meatballs served in a tangy sweet and sour tomato sauce.
Grapefruit Half (in season) 1.25
Chilled Melon (in season) 1.75
Your choice of juice: orange, tomato apple, grapefruit, V-8, or cranberry. .75
Gazpacho 2.25
A chilled, spicy fresh vegetable soup.

ILLUSTRATION 23-3 One aspect of the Dining With Heart program is to highlight healthful meals by including them on restaurant menus with the program's symbol.

Generating Customer Interest

The merchandising campaign for the Dining With Heart program has involved various forms of **media** advertising, such as newspaper, magazine, radio, and television advertisements. In addition, area food writers have been encouraged to write feature articles on the Dining With Heart program, highlighting the participating restaurants. Finally, hospitals have distributed brochures to their medical staffs to give to their patients. These brochures outline the program and list the restaurants providing Dining With Heart menu items. Major health care companies also have indicated a desire to support the program by issuing their own materials. For example, US Healthcare, as part of their Healthy Eating Program, has developed general guidelines for eating out in restaurants while still adhering to special dietary needs. Their Eating Out reference card outlines nutrition guidelines to help restaurant goers with dietary concerns identify healthful items on restaurant menus.[1]

Evaluating Success

The success of the Dining With Heart program cannot be measured on the basis of financial gain. Customer satisfaction is the major gauge of success. The program has not produced a new

[1] Janet Zamsky and Nancy Wolfson, *The US Healthcare Healthy Eating Program* (US Healthcare, 1988), 15–16.

ILLUSTRATION 23-4 US Healthcare's Eating Out reference card outlines nutrition guidelines to help restaurant goers with dietary concerns identify healthful items on restaurant menus.

EATING OUT

Knowing What to Order

It is always easier to eat the right foods when you are eating at home, preparing your own meals. It is more difficult when you are eating in a restaurant. Restaurants don't have to spell disaster. With the help of Healthy Eating, dining out can still be a pleasurable experience. You don't have to panic when you look at a menu. You will be equipped with the knowledge necessary to make the correct food selection. Follow these simple guidelines and you will be fine.

When dining out, whether it's breakfast, lunch or dinner, there is a wide variety of foods to choose from that won't lead you astray from your Healthy Eating habits.

Breakfast

An excellent choice for breakfast is fresh fruit or a small glass of citrus juice (orange or grapefruit).

Order whole grain bread or a dry English muffin. You can use a small quantity of margarine.

Avoid cereals that are high in sugar and sodium. Some prepared cereals are good, but be careful of which ones you select. Order skim or low-fat milk when using milk for your cereal or to drink.

Entrees

When selecting your main course in a restaurant, your best choices include: poultry, fish, shellfish, and vegetable dishes.

These foods are naturally low in fat and can be prepared without adding fat. If you would like to order meat, select lean red meats that are trimmed and prepared according to the Healthy Eating rules. Broiled foods are thought to be low in calories, but even broiled entrees can be basted with fat and seasoned with salt. Order your selection dry broiled or request that juice, wine or a small amount of polyunsaturated oil be used instead of fats.

15

EATING OUT

Entrees

It's best to order food that has been prepared in the following manner:
steamed
in its own juice
broiled
roasted
poached
dry broiled either in lemon juice or wine

Be extra careful NOT to order foods described as:

buttery	creamed
buttered	in cream sauce
in butter	hollandaise
sauteed	au gratin
fried	parmesan
panfried	in cheese sauce
crispy	escalloped
braised	casserole

When ordering salads request dressing on the side, oil and vinegar or lemon wedges.

Side Dishes

When selecting side dishes, such as vegetables and starches, make sure they are cooked fat-free. They should be boiled, baked or steamed, as opposed to fried or cooked in butter and seasoned with salt. When eating baked potatoes, use yogurt rather than sour cream. Stay away from cole slaw and potato salad.

Dessert

If you must have dessert, have fresh fruit.

People who lead active lives and have very busy schedules, don't always have the time to sit down to a leisurely meal. The first thing that usually comes to mind is a fast-food restaurant. That is OK, because many fast-food restaurants have added salad bars. This allows you to control what goes into your salad. It is best to select all fresh vegetables. In addition, they also offer baked potatoes. You should select vegetable or yogurt toppings, or eat them plain. If the fast-food restaurant does not have a salad bar, order a smaller portion of meat (2 ounces of hamburger is acceptable). Avoid fried foods, double-decker burgers, and milk shakes.

16

product but a service to the general public. Although the Dining With Heart menu items have been adjusted in content or presentation, they would have been produced regardless of the program. The Dining With Heart program has however, contributed to the well-being of the community. The common marketing term for this type of activity is public relations. **Public relations** activities foster public goodwill toward a person, firm, or institution. Evaluated in the light of public goodwill and increased awareness of healthful eating, the Dining With Heart program has been, and continues to be, a success.

CAREER PROFILE
ROBERT W. SMITH: VICE PRESIDENT, MARKETING, HARRY M. STEVENS, INC.

Harry M.Stevens INC.

Although a variety of summer jobs had interested Mr. Smith in the hospitality and food service industry, he chose to major in marketing at the University of Bridgeport following high school. During college, the opportunity to own and manage a summer snack bar business in Cape Cod, Massachusetts, challenged Mr. Smith to develop a new restaurant concept, name, and menu. The success of his efforts over the course of two summers resulted in Mr. Smith entering the hotel and restaurant management program at Paul Smith College.

Harry M. Stevens, Inc., a contract feeding company for sports-related facilities, was Mr. Smith's next career stop, and he has been with the company ever since. Starting as a kitchen steward in one of the racetrack dining room kitchens, Mr. Smith soon rotated through a variety of positions. "I'm a hands on person," he says, in reference to his many past jobs. "I learned how to take an idea and make it happen."

Moving to the purchasing department in the corporate offices, Mr. Smith was soon researching needs for new products and following his ideas through every stage of production until they were introduced on the market. Today, Mr. Smith is responsible for marketing food products and service concepts for contract feeding in over 30 sports-related facilities around the United States.

SUMMARY

Marketing, purely defined, is any effort that produces sales. Food service marketing, however, is different from marketing done for other types of businesses. The primary reason for this different viewpoint is the consistent combination of product with service in the food service industry. The major product of any food service operation, food, cannot be delivered without the second major product, service. It is impossible to separate the two when developing a food service marketing campaign. Likewise, the other elements of the traditional marketing mix—place, promotion, and price—become presentation and communication in the food service marketing mix, reflecting the importance that the industry places on atmosphere and customer service.

Merchandising is the promotion or advertisement of a product-service. Any form of communication that can inform a customer of the availability, content, location, and price of a product-service is considered merchandising. In the food service industry, merchandising techniques range from newspaper advertisements to restaurant menus.

Going through the four-step marketing cycle is the most effective way to ensure the success of a product-service. This formal plan serves as an established route for identifying a need, developing a product-service, generating customer interest, and evaluating success.

Marketing and merchandising are an exciting, creative part of the food service industry. They represent some of the many career alternatives offered by the food service and hospitality industry.

QUESTIONS

1. What is the concept of marketing?
2. What does the term *media* mean?
3. What is the path of the marketing cycle?
4. What is the difference between identifying a need and creating a need (as defined in Chapter 19)?
5. How can one create customer interest? Give an example.
6. By what two criteria can customer satisfaction be measured? Give an example of each.
7. What does the term *merchandising* mean?
8. What steps should management take before putting any marketing concept into action?
9. What are the four elements of the traditional marketing mix?
10. *Public relations* is a common marketing term that reflects what kind of marketing activities?

ACTIVITIES

Activity One With the help of your teacher, create a food service marketing campaign. Using your classmates as your customer base, identify or create a need to which they will respond.

Activity Two Design a restaurant promotion for in-house use. Create a poster or table tent for the promotion.

Activity Three Using your area newspapers and magazines, review the restaurant advertising in your community. What types of promotions are being offered? What types of marketing campaigns can you identify in the advertisements?

Glossary

Abstract The cost of any food or seasoning not included in a recipe but used in the service of an item, such as a condiment. Two percent of the total portion cost is the average figure used for an abstract.

Accent copy Text that creatively highlights menu items, often by drawing attention to their names. *See also* Descriptive copy; Merchandising copy.

Accounts payable A record of bills that must be paid, determined from the invoices and handled by the accounting department of an operation.

Accounts receivable A record of accounts that are owed to an organization, determined from the information on guest checks and handled by the accounting department.

À la carte A type of menu that prices each item separately. *See also* Table d'hôte.

Aluminized A term describing metals (or other materials) that have been treated or coated with aluminum. Steel cooking equipment is usually aluminized or copperplated.

American nouvelle cuisine Literally, the new American cuisine. Refers to menu item variations developed by American chefs today in an attempt to enhance traditional menu items.

American plan A hotel payment plan in which all three meals are included in the price of accommodations. *See also* European plan.

American service A style of table service in which all food is cooked and arranged on china in the kitchen rather than at the table side; it dramatically reduces the number of servers needed in the dining room. *See also* Buffet service; Family style service; French service; Russian service.

Ampere A unit of electrical current equivalent to the current produced by one volt flowing through a conductor whose resistance is one ohm.

Anthocyanins Natural red or purple pigments found in fruits and vegetables such as red cabbage, beets, cherries, and plums.

Anthoxanthins Natural white or colorless pigments found in vegetables such as potatoes, onions, cauliflower, and turnips.

Aperitif An alcoholic drink taken before a meal as an appetizer; a cocktail.

Appetizer A predinner food or drink that stimulates the appetite.

Apprentice A person who is learning, by practical experience under a skilled worker, a trade, art, or calling.

Apprentice cook *(commis)* An apprentice who is training to become a chef by working as a kitchen assistant.

Arteriosclerosis A disease characterized by fatty deposits, or plaque, on the arterial walls. Plaque can block the flow of blood, causing heart attacks or strokes.

Attitude A feeling or emotion about something or someone. In food service, personal and work attitudes are important because they directly affect an employee's productivity and the attitudes of customers and other employees.

Back-of-the-house position In a restaurant, any job that involves working outside of public space. *See also* Front-of-the-house position.

Bacteria Microscopic plants, some of which are harmful to humans, plants, and animals. *See also Clostridium botulinum; Clostridium perfringens; Salmonella; Staphylococcus aureus.*

Banquet service A service style used for groups of customers, requiring specialized forms of table service (called Russian service) and seating arrangement. *See also* Cafeteria service; Family style service; Fast-food service; Full service; Take-out service.

Barding A process in which fat is placed on top of a lean piece of meat and allowed to soak through it during roasting, making it more tender. *See also* Larding.

Beurre manié A mixture made with equal amounts of flour and butter and then kneaded together; used to thicken thin sauces. *See also* Roux.

Béchamel sauce A basic milk sauce made with roux and hot milk. This base is mixed with other ingredients to make a variety of sauces, such as Newburg and cheese sauce. *See also* Roux.

Blending A mixing technique that involves combining two or more ingredients thoroughly with a spoon.

Borscht A beet soup with diced onions, sugar, and vinegar or lemon juice. Served hot or cold, often with sour cream.

Bouillon Broth that is made from meat, poultry, or fish stock, with all visible fat skimmed off.

Bran The outermost layer and protective coating of a grain. Bran contains cellulose, some protein, thiamin, and minerals. It is a grain's best source of fiber. *See also* Endosperm; Germ.

Braziers Large kettles for cooking with small amounts of liquid.

Brigade de cuisine *See* Kitchen brigade.

Brigade de service *See* Dining room brigade.

Broilers Cooking units that directly expose food to dry heat from charcoal, gas flames, or electric metal coils.

Brunch The combination of breakfast and luncheon items. Brunch usually includes fruits, eggs, fish, meat, breads, desserts, and beverages.

Buffet service A style of table service that involves setting out a variety of food items from which guests can help themselves. This type of service significantly reduces labor expenses. *See also* American service; Family style service; French service; Russian service.

Bus person *(commis debarrasseur)* In a full-service restaurant, a person who cleans and sets up the tables, and who often assists in serving.

Cafeteria service A service style usually associated with institutional and contract feeding (which involves an outside service that is contracted to run a food service operation). *See also* Banquet service; Family style service; Fast-food service; Full service; Take-out service.

Calorie The amount of heat required to raise one kilogram of water one degree Celsius. Calories measure body energy.

Carbon dioxide (CO_2) A colorless, odorless gas that makes dough rise. CO_2 is produced when yeast is added to baking products.

Carbon dioxide extinguishers Fire extinguishers that smother a fire without conducting electricity. CO_2 gas is extremely cold and can lower a fire's heat level without damaging equipment.

Cardiopulmonary resuscitation (CPR) A first aid technique used to treat patients whose heart and breathing have stopped. Often used on heart attack victims.

Carotenoids Natural orange, red-orange, yellow, and red pigments found in vegetables such as carrots and squash.

Central processing unit (CPU) A piece of computer equipment that stores data provided by the POS system and, with the keyboard and printer, generates operational records, spreadsheets, and other accounting records.

Cents factor (CF) The percentage of the extension total that represents any small costs for ingredients that are included on a cost card, such as a pinch of spice or a tablespoon of flavoring.

CF *See* Cents factor.

Chafing dish A serving dish used in buffet service to keep hot food hot with a source of liquid heat, such as Sterno. The four parts of a chafing dish are the stand, the water pan, the insert pan, and the cover.

Chlorophyll A natural green pigment found in vegetables such as broccoli, beans, and peas.

Circuit breaker A device that automatically interrupts an electrical current when the flow of energy becomes excessive.

Clarification The removal of sediments; clarification produces a clear soup, such as consommé, that can be decanted.

Class A fires Fires involving combustible materials, such as wood, paper, and cloth. A Class A fire can be extinguished by the removal of its fuel or the cooling of its heat.

Class B fires Fires fueled by grease or by oil. A Class B fire can be smothered by removing its oxygen supply.

Class C fires Electrical fires. A Class C fire can be smothered by removing its oxygen supply.

Clostridium botulinum A bacterium that affects the nervous system and causes botulism, a rare but lethal illness. Improperly processed, non-acidic canned foods provide an excellent medium for this bacterium to grow.

Clostridium perfringens A spore-forming bacterium that thrives in the absence of oxygen. Foods susceptible to this bacterium include raw meats, meat-based gravies and sauces, and cooked meat and poultry left at room temperature for several hours.

Commercial restaurant Any restaurant that exists to make a profit for owners and investors. *See also* Institutional food service; Nonprofit business.

Commis *See* Apprentice cook.

Commis debarrasseur *See* bus person.

Commis de range *See* principal server.

Communication The process of relaying and receiving information. Good communication skills are important for a server's and a manager's success.

Compensation Employee salaries, wages, and benefits. Labor costs, in the form of compensation, make up a substantial part of an operation's total costs.

Competition Any businesses that directly compete for an organization's customers.

Concept restaurant *See* theme or concept restaurant.

Condiment An item used to make food more savory, such as catsup, mustard, or relishes.

Consommé A concentrated soup made from the clear broth of meat, poultry, or fish plus other ingredients, such as lean ground beef, ground vegetables, egg whites, or ice water. After all the ingredients are combined, the soup is clarified. *See also* Clarification.

Continental breakfast An informal breakfast that consists of fruit or fruit juices, breakfast breads, and beverages.

Contract A written agreement between a buyer and a vendor, with the signatures of both parties. A well-written contract should include the offer, the vendor's acceptance, and the price, time, and date of delivery.

Convection oven An oven with a fan that circulates hot air to speed the cooking process without drying out food. It cooks food more quickly at lower temperatures than a conventional oven.

Conventional oven An oven that cooks food by heating air in an enclosed space. It is best for baking cakes and other foods made with batter.

Cook-chill technology A food manufacturing technology that involves prepackaged, portioned convenience foods that are refrigerated and can be prepared for eating with minimal effort.

Cook-freeze technology A food manufacturing technology that involves prepackaged, frozen convenience foods that can be prepared for eating with minimal effort.

Copperplated A term describing metals that have had copper bonded to them.

Cordials Drinks that are served after a meal to aid digestion.

Cost card A form that outlines the cost of each ingredient in a recipe. When all of the costs are totaled, the cost of each portion of a recipe can be determined. *See also* Recipe card.

Cost of purchased unit (CPU) The price that an organization pays for a unit of food (a unit is the measure or weight in which an item comes at the time of purchase).

Courses Groups of food that are served in an established sequence during a meal. In the 1800s, the French established the seven-course menu format upon which modern menus are based.

Courtesy Polished manners and helpful, considerate acts or remarks based on a respect for other people. Courtesy is an important aspect of a server's job.

CPR *See* Cardiopulmonary resuscitation.

CPU *See* Central processing unit; Cost of purchased unit.

Creaming A mixing technique that involves incorporating air into a mixture to make it smooth and creamy.

Creative imitation A theory of entrepreneurship stating that a new consumer market can be created by starting with a successful product or service and doing something different to it. The end result is a new product or service with which the consumer is already familiar. *See also* Innovation; Leadership.

Cuisine A manner of preparing food; a style of cooking. Cajun cooking, Tex-Mex, and California nouvelle are examples of different cuisines.

Customer profile A series of general characteristics that identifies the target market to which a customer belongs. Age, occupation, geographical location, and annual income are characteristics that make up a typical customer profile.

Cutting in A mixing technique that involves blending two ingredients together by rubbing them with the palms of the hands or by using a pastry cutter.

Daily food cost sheet A record of the total cost of food requisitions balanced against the total sales for a day; produces a food cost and a food cost percentage.

Daily server report A record of all of a server's transactions for a day. Information for this report comes from guest checks.

Daily server sales report A record of the transactions for all of the servers working in a restaurant on a particular day; allows management to quickly review the sales totals for all servers.

Deep-fat fryer A piece of equipment designed to cook food in hot fat. Deep-fat fryers can cook foods to a desired consistency and crispness without making them greasy; many of them are now computerized.

Deglassing Adding water to a pan in which meat has been roasted, sautéed, or pan broiled to dissolve the juices that have dried on the bottom and sides of the pan. The juices can then be used to make gravies or sauces.

Delegation The act of giving another person the responsibility for a job and the permission to perform it.

Delicatessen A shop that sells ready-to-eat foods, such as cold meats, salads, and sandwiches.

Descriptive copy Text that explains a menu item's ingredients, cooking style, and presentation, usually including many descriptive words to appeal to a customer's senses. *See also* Accent copy; Merchandising copy.

Deuce In a restaurant, a table with two seats.

Dietitian A specialist in dietetics (the science of applying nutrition principles to the diet) responsible for menu planning and development in a food service establishment.

Dining room brigade *(brigade de service)* A team of front-of-the-house employees who perform all the service tasks in a restaurant.

Diplomacy The art of dealing tactfully with people, an important service skill.

Dry heat cooking A method of cooking in which heat is transferred by radiant energy and air. Energy sources for dry heat cooking include electric heating elements, gas flames, hot coals, and flames from wood and charcoal. *See also* Moist heat cooking.

Earthenware Ceramic tableware made of slightly porous opaque clay fired at low heat. Often called pottery. Many of the first cups and plates were made of earthenware.

Electronic cash register A sophisticated cash register that issues customer receipts and sales records.

Emulsifying agent An ingredient that helps combine two other ingredients that do not mix well, such as oil and vinegar. Eggs are often used as emulsifying agents.

Endosperm The softest component of a grain; contains all of a grain's starch. *See also* Bran; Germ.

Enriched A term describing flour in which the iron, niacin, thiamin, and riboflavin lost through the milling process are put back.

Entrées Main course items. In Russian service, entrées are placed on platters; in French service, they can be prepared at the table; in American service, they are plated in the kitchen.

Entrepreneur A person that manages and assumes the risks of a business through the application of basic business management principles.

Enzymatic browning A process that occurs when the enzymes in a piece of cut fruit react with oxygen, causing the fruit's flesh to turn brown. Enzymatic browning can be prevented by cutting fruit just before serving or by dipping cut fruit in an acidic solution, such as lemon juice and water.

Escargot A snail prepared as food. Specialty utensils are used to eat escargot.

Escoffier dishes Dishes with two compartments, used in Russian service to serve vegetables and starches.

European plan A hotel payment plan in which meals are billed separately. *See also* American plan.

Evaporated milk A sterilized milk with half its water removed; can be used full strength or as a cream on foods. Evaporated milk is canned and may be stored a long time without refrigeration.

Exploded A term describing increased amounts of ingredients in a recipe. If more portions of a certain recipe are needed, quantities of ingredients are exploded as necessary.

Extension The cost of the amount of an ingredient used in a recipe, calculated by multiplying or breaking down the CPU (Cost of purchased unit) according to the ingredient's weight or measure.

Fair Labor Standards Act of 1938 An act that established overtime and minimum age requirements. The act requires employers to pay employees time and a half for all hours worked in excess of 40 hours per week and also stipulates that all employees must be at least 16 years old and at least 18 years old for hazardous jobs.

Fair Packaging and Labeling Act of 1967 An act that requires food manufacturers to provide customers with accurate label information describing a food's quality and content. A manufacturer must include its name and address, the name of the food, the net content of the food, and the name of all the ingredients listed in order of their percentage of the food.

Family style service A service style (and type of table service) that allows diners to join others in a setting where large dishes of food are placed on a table for self-service. Often called English service. *See also* American service; Banquet service; Buffet service; Cafeteria service; Fast-food service;

French service; Full service; Russian service; Take-out service.

Fast-food service An operational style utilizing efficient preparation and prepackaging techniques to produce foods for consumption at home or in the restaurant. *See also* Banquet service; Cafeteria service; Family style service; Full service; Take-out service.

FC% *See* Food cost percentage.

FDA *See* Food and Drug Administration.

Federal Trade Commission (FTC) A government agency that is responsible for dealing with false claims made by any company, such as claims concerning nutritional content.

Federal Wage Garnishment Act An act that protects an employee's income by limiting the amount of wages that may be garnished, or withheld, when his or her employer is in debt.

Fermentation The process by which yeast, feeding on starch and sugar, produces CO_2 gas, thus increasing a dough's volume.

FIFO *See* First in, first out.

First in, first out (FIFO) An important principle of storage stating that old and new stock should be shelved so that older stock is used first.

File number A number on a recipe card that refers to the recipe's place in an organization's total file system. File numbers are assigned first by menu category and then by number.

Fire extinguisher A device used to put out a fire. Different types of fire extinguishers are used for different classes of fires.

Fixed bid A method of purchasing that consists of providing written specifications for an item that is needed in large quantities for a long period of time. The price on a fixed bid remains the same for the period agreed upon. *See also* Weekly bid.

Flatware The utensils used to serve and eat food, made out of a variety of metals or disposable plastic.

Fluorescent light A type of light produced in glass tubes coated on the inside with a substance that gives off light. Fluorescent lights are available in a wide range of warm colors. *See also* Incandescent light.

Folding A mixing technique that involves gently rotating a spoon or spatula from the top of a mixture to the bottom and from the bottom back up, as in folding beaten egg whites into a batter.

Food and Drug Administration (FDA) A government agency that is responsible for enforcing the Food, Drug, and Cosmetic Act of 1938, which regulates the production, manufacturing, and distribution of some foods involved in interstate commerce. The FDA also checks whether food labels are accurate and include required information.

Foodborne infection An illness caused by eating foods containing harmful bacteria or parasites, which affect an individual's gastrointestinal tract.

Foodborne intoxication An illness caused by toxins that have formed in a food prior to consumption.

Food cost The cost of any food or food-related items used in an operation, including nonalcoholic beverages. The food cost for a particular dish is the total portion cost as calculated on a cost card.

Food costing The process of determining the cost of a recipe's ingredients.

Food cost percentage (FC%) A food item's cost to a business expressed as a percentage of the item's selling price. To ensure a profit, a manager must make sure that an item's cost represents an appropriate percentage of the selling price.

Food halls Centralized areas for a variety of fast-food restaurants and seating, usually in shopping malls and downtown areas.

Food mixer A piece of equipment that can perform a range of functions: it can mix, beat, knead, whip, slice, strain, chop, grind, heat, and cool mixtures.

Food service management The coordination of people, resources, products, and facilities related to the design, preparation, and presentation of food outside of the home.

Food service marketer A person who is involved in the creation, purchase, and distribution of products for the food service industry.

Food slicer A piece of equipment that cuts or slices foods to a consistent thickness. Larger models may be motor driven, while smaller models are operated manually.

Food stylist A person who designs plate and food presentations for advertising and public relations photographs and films.

Food writer A writer whose major focus is food. Food writers can write about recipes, restaurants, or nutritional issues.

Forecasting The ability to estimate the what, where, how, and how much of future product purchasing, based on previous experience and current knowledge. Purchasers must be able to forecast future needs based on market and economic conditions.

French service The style of table service around which most other table service styles have evolved; can involve up to six members of the dining room staff, who form a *brigade de service* (dining room brigade). French service is characterized by table-side cooking. *See also* American service; Buffet service; Family style service; Russian service.

Front-of-the-house position In a restaurant, any job that involves guest service. *See also* Back-of-the-house position.

Fructose A natural form of sugar that gives fruit its sweetness.

FTC *See* Federal Trade Commission.

Full service A service style that provides a sit-down dining room and table service. Some full-service restaurants also provide cocktail lounges and entertainment. *See also* Banquet service; Cafeteria service; Family style service; Fast-food service; Take-out service.

Galvanize To coat with zinc. When galvanized containers are used to cook or prepare high-acid foods, metal poisoning can result.

Gazpacho A highly seasoned tomato and vegetable soup made from beef stock, served chilled with finely chopped cucumbers and tomatoes.

Geridon A rolling cart used in table-side food preparation; can be used for salads, fish, or meat entrées, and for flambéed desserts.

Germ The innermost layer of a grain; rich in protein, fat, and B-complex vitamins. The germ is the part of a grain that reproduces, and thus it contains concentrated amounts of nutrients to support new growth. *See also* Bran; Endosperm.

Gluten An elastic protein that makes dough cohesive.

Gratuities Tips left in return for service, usually based on the total check amount.

Grounded equipment Equipment with a three-pronged plug that allows electricity to travel to and connect with the ground rather than with the person handling the equipment.

Hard copy The printed record of a computer-generated transaction, such as a receipt or a check.

Health inspectors Government officials who inspect food service operations periodically to ensure that they meet the health and safety regulations set by the federal, state, and local governments. In addition to inspecting for the sanitary handling of food, they check to see that an establishment has eliminated safety hazards.

Heimlich maneuver A first aid procedure designed to help choking victims. To perform the maneuver, apply pressure below the victim's rib cage to expel food clogged in the esophagus.

Hollandaise sauce A basic egg sauce made with egg yolks and butter. Commonly served over vegetables, it is also a base for many other sauces.

Homogenization A process in which the fat globules and fat-soluble vitamin A in raw milk are broken into small particles and dispersed throughout the product, resulting in a more uniformly textured milk. All milk sold for public consumption must be homogenized. *See also* Pasteurization.

Hygiene The science of health maintenance through cleanliness and the use of effective sanitary procedures.

Incandescent light A type of light produced by a filament of conducting material contained in a vacuum and heated by an electrical current. Yellow-white incandescent lights are preferred for food displays placed in a dining area. *See also* Fluorescent light.

Inn An establishment for lodging, entertaining, and serving meals and alcoholic beverages to travelers.

Innovation A theory of entrepreneurship that, when applied, creates potential customers by identifying a need, determining what product or service the customers want, and establishing how much they are willing to pay. *See also* Creative imitation; Leadership.

Institutional food service An organization that serves meals and à la carte items for a larger, often publicly-owned facility, such as a hospital, school, retirement home, or prison. *See also* Commercial restaurant; Nonprofit business.

Intoxicated A term describing a person who is affected by alcohol to the point where physical and mental control is markedly diminished.

Inventory A list of the supplies and food items on hand, detailing all the items that a business uses.

Invoice A record-keeping form used to list all goods shipped, specifying their prices and the terms of the sale. When requested items are received by a buyer, the invoice should be compared to the purchase order for accuracy.

Job description A detailed outline of the duties of a certain job; along with a job specification, this description allows both employers and applicants to fully understand a position.

Job orientation A program designed to acquaint a new employee with her or his job. Orientation should include the history and goals of the organization; its policies, procedures, benefits, employee expectations; introductions to other employees; and a tour of the facility.

Job specification A detailed outline of the skills and qualifications needed to do a specific job. Along with a job description, this outline allows employers and applicants to fully understand a position.

Keyboard The set of systematically arranged keys on a cash register. Keyboards can be touch sensitive and color coded to help servers distinguish between menu categories.

Kitchen brigade *(brigade de cuisine)* A team that performs all the food production tasks for a restaurant under the direction of the executive chef *(chef de cuisine)*.

Labor In relation to food cost math, labor is the direct cost of employing help, as in hourly wages and salaries. It also includes the indirect costs of employee benefits, insurance, Social Security, and training.

Lactase An enzyme found in the human intestine that helps metabolize milk sugar (lactose). *See also Lactobacillus acidophilus;* Sweet acidophilus.

Lactobacillus acidophilus A bacterium that, when added to milk, separates the milk sugar (lactose) into glucose and galactose for people who lack the enzyme lactase and thus cannot properly digest milk. *See also* Lactase; Sweet acidophilus.

Larding A process in which strips of fat are pushed into the inner part of a lean piece of meat and allowed to soak through it during roasting. Larding makes a lean piece of meat more tender. *See also* Barding.

Layout A plan that shows a facility's physical setup, including construction features and the locations and space allowances for different work stations and equipment.

Leaders In food service, items on a menu that have high sales volumes. *See also* Losers.

Leadership An entrepreneurial theory that an entrepreneur applies when developing a product or service that has never before existed. *See also* Creative imitation; Innovation.

Leavening agent A substance that helps baking products rise; air, steam, yeast, and chemical leaveners are the most common leavening agents.

Linens The fabrics used on a tabletop. Linen colors, patterns, and styles are a major part of a restaurant's theme.

Losers In food service, items on a menu that have low sales volumes. *See also* Leaders.

MADD *See* Mothers Against Drunk Driving.

Main course Traditionally the fourth course of a meal. Public dining habits have changed the main course, or entrée, into the second course, following the appetizer.

Maîtres d'hôtel Dining room managers. After the French Revolution, many maîtres d'hôtel left France for other parts of Europe, taking with them their traditions of table service.

Marbling The visible fat that is embedded in the lean tissue of meat. The amount of marbling depends on the age of the animal, the type of feed, and the amount of exercise the animal received.

Market research survey An examination of a specific group of customers, competitors, and target markets within an established geographical area; helps an entrepreneur know who his or her customers are and what these customers want.

Marketing Business activities that direct the flow of goods and services from producers to consumers and thus affect sales.

Marketing cycle A series of steps that must be followed to successfully market a product or service. The steps are identifying a need, developing a product or service, generating customer interest, and evaluating financial success and customer satisfaction.

Marketing mix The marketing elements concerned with customers' needs: product (or service), place, promotion, and price. The specialized food service marketing mix consists of product-service, presentation, and communication.

Masonry Building materials made of brickwork or stonework.

Media Agencies of mass communication, such as newspapers, magazines, radio, and television.

Memory Capacity A term that refers to a computer's ability to store data. An electronic cash register has limited memory capacity, whereas a central processing unit has a large memory capacity.

Menu A printed form that outlines all of a restaurant's offerings, usually organized by a series of courses.

Menu copy Text that is used to introduce and describe the menu items to a customer. Creative menu copy can increase sales by raising customer interest in different foods. *See also* Accent copy; Descriptive copy; Merchandising copy.

Menu layout The way in which illustrations, graphics, and the meal course structure are arranged on a menu. This visual presentation of menu items can determine what choices a customer makes.

Merchandising Sales activities that generate customer interest. These activities include a wide range of promotional items or events as well as advertisements in magazines or on radio and television.

Merchandising copy Text in a menu that is meant to create interest in a restaurant, including historical information, profiles of the owners, or a discussion of the featured cuisine. *See also* Accent copy; Descriptive copy.

Microwave oven An oven that cooks food with microwave, or radiant, energy. A microwave oven converts electricity into microwaves, which bounce around the interior of the oven and enter food from all sides. It can prepare food in approximately one-fourth the time of a conventional oven.

Milligram A metric measurement that is equal to one-thousandth of a gram.

Modular computer system A computer system

that consists of a number of separate stations that can operate independently of one another or in unison. One station, for example, could be behind a restaurant's bar, one at the manager's workstation, and one at the server's workstation.

Moist heat cooking A method of cooking requiring moist air and water. Moist heat cooking techniques include boiling, braising, stewing, steaming, and pressure cooking. *See also* Dry heat cooking.

Monitor One of the three pieces of computer equipment in a POS system, together with a cash drawer and keyboard. The monitor is a screen on which information is displayed.

Monosodium glutamate (MSG) A commonly used flavor enhancer that can cause illness if consumed in excess.

Mothers Against Drunk Driving (MADD) An organization of parents that promotes alcohol awareness and training programs in an attempt to reduce the number of injuries and deaths involving drunk drivers. *See also* Students Against Drunk Driving.

Motivation The need or desire to perform a specific task; usually involves setting and reaching goals, and is strongly related to a healthy attitude.

MSG *See* Monosodium glutamate.

Noncompliance The act or process of not fulfilling official requirements; in the case of safety regulations, it can result in penalties that range from a $10,000 fine to six months in jail for an employer.

Nonprofit business A business that provides a service to the community, such as a private club with a limited membership. *See also* Commercial restaurant; Institutional food service.

Occupational Safety and Health Administration (OSHA) An organization created within the Department of Labor that attempts to keep places of employment free from known health hazards.

Order sheet A record-keeping form listing all of the items that have been requested but are not in inventory, broken down by category.

Ordinary An establishment providing lodging and regularly scheduled meals.

Organizational chart An outline of all of the job categories within an organization. Jobs are placed from top to bottom in order of decreasing responsibility.

OSHA *See* Occupational Safety and Health Administration.

Osteoporosis A disease characterized by the degeneration of bone tissue and caused by a calcium deficiency. Osteoporosis is more common among women than men.

Oven An insulated heating unit of various capacities designed to cook food in covered or uncovered baking pans. *See also* Convection oven; Conventional oven; Microwave oven; Revolving oven.

Overhead The operating expenses of a business, such as rent, taxes, utilities, and insurance.

Oxidized A term describing food containing fat that has been overexposed to oxygen. Oxidation can cause food to go rancid, or to develop offensive odors and flavors.

Parasite An organism living in or on another organism. Some parasites can cause foodborne illnesses. *See also Trichinella spiralis.*

Parboiled A term describing food that has been boiled briefly as a preliminary or incomplete cooking procedure.

Pasteurization A partial sterilization process in which raw milk is heated at 165°F for 15 minutes. This process kills harmful bacteria that are naturally present in raw milk. All milk sold for public consumption must be pasteurized. *See also* Homogenization.

***Pastilliage* decorations** Elaborate presentation or show pieces prepared by a pastry chef, made of a hard sugar paste.

PC *See* Portion cost.

Perceived value The amount that buyers feel a product or service is worth. Customers usually will not pay more than the perceived value.

Percentage of yield A figure indicating the waste factor created by returned items in a food service operation, as calculated on various food cost reports.

Performance evaluation A written record that provides employees with feedback on the acceptability of their work, encourages and praises employees for a job well done, and documents information to help employers make future decisions about job assignments and promotions.

Perpetual inventory A record-keeping form on which employees can record the receipt, withdrawal, amount, and cost of items in cold and dry storage areas.

Pesticides Chemicals used to kill pests during plant growth; can cause food contamination.

pH A measure used to indicate the acidity or alkalinity of an object or solution. Acidic mediums have a pH less than seven, and alkaline mediums have a pH greater than seven. Changing a fruit or vegetable's pH level can change its color.

Pheophytin A substance developed when the acids in green vegetables remove a magnesium ion from chlorophyll, causing the vegetables to change color from bright green to olive green or grayish green. Cooking in tap water that is slightly alkaline can prevent these color changes.

Physical inventory A monthly or quarterly report on all the items in a food service facility's storage areas.

Pigments Substances that give color to other materials. The four natural pigments in vegetables are chlorophyll, anthocyanins, anthoxanthins, and carotenoids. *See also* Anthocyanins; Anthoxanthins; Carotenoids; Chlorophyll.

Place setting China, glassware, and utensils that have been selected to enhance a restaurant's tabletop design and operational style.

Polygraph Protection Act An act stating that restaurant operators and other employers cannot use lie detectors to screen job applicants. The act also prohibits an employer from asking an employee to take a polygraph test unless the employer is investigating a case of theft.

Porcelain The finest base for food service china; made from a base of clay, feldspar, quartz, and, often, bone.

Porringer A low metal bowl with a handle. Early Colonial settlers brought pewter porringers from England.

Portion The amount of the finished product of a recipe that is served to a customer.

Portion cost (PC) The cost per portion (to a food service facility) of a menu item, calculated by dividing the extension total (plus the cents factor) by the number of portions yielded by a recipe.

Portion size The amount of a food item served per order. Most restaurants offer a range of portion sizes for their entrées.

POS (point-of-sale) system A computer system that begins at the point of sale and helps facilitate efficient food service.

Potassium bromate A chemical used to make flour more stretchy and pliable for easy handling. Bromated flour is good for making bread products.

Pounding A process in which a less expensive, lean cut of meat is pounded with a wooden or metal mallet before pan frying. This process shortens its fibers and produces a more tender product.

Preservatives Chemicals used on foods to maintain freshness and prevent spoilage; can cause contamination of foods. *See also* Monosodium glutamate; Sodium nitrite.

Pressure steamer Equipment that cooks by directing 5 to 15 pounds of pressurized steam toward food.

Principle server *(commis de range)* The server who presents the menu, explains any additions, and takes any predinner beverage orders.

Product developer A person who refines or creates ways to prepare, serve, and package food products.

Production sheet A form used to forecast, or predict, the number of each menu item that will be produced for a given meal service on a given day. This form serves as the basis on which a chef requisitions food supplies and schedules kitchen personnel.

Profit A business's monetary gain after all expenses are paid, including taxes.

Promotions Activities developed and conducted to increase the growth of a business or the desire for a service.

Proofing A term describing the final rising of a dough, after shaping but before baking. This process requires a warmer temperature than fermentation.

Public relations The business of cultivating public goodwill toward a person, firm, or institution.

Purchase order A record-keeping form used during the purchasing process to state specific information regarding the items to be purchased. It also serves as an agreement between a buyer and a vendor.

Purchase requisition A record-keeping form used within an organization to request food and supply items.

Purchasing The act of acquiring goods for a given price; buying.

Purveyor A supplier who identifies a customer's needs, develops sales strategies, and distributes products.

Radiant energy Energy that travels in a wave motion; used in microwave ovens.

Rancidity A condition characterized by the development of offensive odors and flavors when the fat in a food has been overexposed to oxygen, or oxidized.

Range The flat top of an oven with gas burners or electric heating units; designed mainly for surface cooking and the heating of food in pots and pans.

Rechaud A small, open-flame heating unit that is placed on a *geridon* (a rolling cart) and used in table-side cooking.

Recipe A set of written instructions for making a food item from specific ingredients. Like a chemistry formula, when a recipe is followed exactly a consistent product results.

Recipe card A detailed outline of the ingredients and directions needed to produce a food item, usually typed on heavy-paper index cards. A recipe card should include a file number and information on yield and portion size. *See also* Cost card.

Reconcile To check an invoice against an order, the last step (before payment) in the purchasing process.

Remote printers Computerized printers located in areas away from the keyboard, the central processing unit, and the monitor in a POS system.

Restaurant A public eating place. The word restaurant was derived from the French word *restaurer,* which means "to restore" or "to refresh."

Retail A term describing the form of commodities or goods that are sold in small quantities to the ultimate consumer; the form of an item when it is sold in a grocery store.

Retirement Income Security Act of 1974 An act that protects employee pension plans by restricting employers' rights to operate and control them. Also called the Pension Reform Law.

Revenue per cover The average amount of money spent by each customer, as calculated on a sales report.

Revolving oven An oven used to bake and roast meats in large quantities, such as for large school districts and bakeries. Shelves inside a revolving oven are built ferris-wheel style and are turned by a gear-driven mechanism outside the heating area.

Roll printer A small, computerized printer used in a restaurant as a remote unit. Prints on a roll of paper similar to the kind used in a cash register and appropriate for recording orders.

Rounding A mixing technique that involves shaping dough on a floured surface to seal its ends.

Roux A mixture of flour and fat in equal proportions. The fat used may be butter, margarine, oil, chicken fat, or bacon fat; the flour can be from rice or wheat. Roux is added to soups and sauces as a thickening agent. *See also* Beurre manié.

Russian service A style of table service that is

generally used for banquets that involve table seating for 6 to 12 guests. All food is fully prepared in the kitchen and served by portions. *See also* American service; Buffet service; Family style service; French service.

SADD *See* Students Against Drunk Driving.

Salary Money earned on a fixed basis, usually by people in management and professional positions. *See also* Wages.

Sales history form The daily journal of a restaurant; records special events or problems that may have affected business on that particular day. The number of reservations, the total number of customers served at each meal service, and the total revenue are also noted.

Sales mix record A form that shows the amount of each menu item sold over a specified period of time, usually a month. Provides a basis for forecasting production schedules and volume purchases.

Salmonella A rod-shaped organism that can cause food poisoning or gastrointestinal inflammation. Found in a wide range of foods, *Salmonella* bacteria are easily destroyed by heat.

Sanitary engineer A person who inspects food service operations periodically to ensure that they meet the sanitation guidelines set by federal, state, and local governments.

Sanitation The promotion of hygiene using practices that keep food free from disease-causing microorganisms.

Sauce A thickened, flavorful liquid that enhances the taste of prepared foods. Sauces are classified as warm, cold, or sweet, and they are prepared from a variety of ingredients.

Sautéing Frying vegetables or meats in a small amount of oil or fat. Food should be cut into small pieces when sautéing to ensure rapid and uniform cooking.

Scurvy An illness caused by a deficiency of Vitamin C, which is abundant in citrus fruits. Scurvy weakens and causes pain in the muscles and joints.

Selling price The price at which an item is sold; should cover all costs and ensure a profit.

Semistructured interview An interview in which only some of the questions are prepared ahead of time. This type of interview is more flexible for an interviewer because it allows him or her to ask additional questions. *See also* Structured interview; Unstructured interview.

Semolina A high-protein durum wheat product used to make quality pasta.

Serviceware Any item used in front-of-the-house areas to serve guests; usually includes flatware, china, and glassware.

Serviette A silk napkin.

Show plate A plate used for display purposes only. The actual dinner plate is set down following the removal of the appetizer course and just before the entrée is served.

Side station A front-of-the-house work area, located in the dining room of a restaurant, where servers can store condiments, extra tableware, napkins, and cleaning equipment. Can also be used as a water and coffee station.

Sifting A mixing technique that involves passing dry ingredients through a sieve or flour sifter to make a lighter product.

Sneeze guard A plastic or glass barrier over a salad bar that helps prevent foods from being contaminated.

Social Security Act of 1935 An act that protects employees against the loss of earnings after retirement, unemployment, disability, or the death of a spouse. The program is supported by a tax withheld from employees and an amount paid by employers.

Sodium A chemical element that is needed to maintain proper water balance in the body. Too much sodium can cause water retention, swelling, and high blood pressure.

Sodium nitrite An additive used in lunch meats, bacon, hot dogs, and other products. The long-term effects of eating foods containing sodium nitrite are unclear, although some experts believe that overconsumption is harmful.

Sommelier *See* Wine steward.

Specification A statement describing the quality, characteristics, and quantity of a product requested for purchase, including allowable variations. A specification should include the name brand, federal grade, size, weight, and unit price on which the final price is established.

Staphylococcus aureus A pathogenic bacterium that is found on the human body and can be passed to meats, poultry, eggs, fish, shellfish, and other protein-rich foods. Commonly called staph.

Steam-jacketed kettle A kettle made of two stainless steel containers, one inside the other. Steam-jacketed kettles have interior capacities of more than two hundred gallons and cook using steam pressure.

Stir frying A cooking process that involves frying vegetables, meats, poultry, or fish with hot oil in a heavy metal pan called a wok. Food is stir-fried at high temperatures for a short period of time.

Stock A liquid ingredient produced by simmering meat, fish, or poultry bones and scraps for several hours in water until their flavor and nutrients have been extracted and the water has become concentrated. Stock is the base ingredient in soups, sauces, and gravies.

Structured interview An interview in which an interviewer asks all applicants the same questions, prepared ahead of time, and records each applicant's responses, thus minimizing possible bias or prejudice by the interviewer. *See also* Semistructured interview; Unstructured interview.

Students Against Drunk Driving (SADD) A student organization that strives to promote alcohol awareness and training programs in an attempt to reduce the number of injuries and deaths involving drunk drivers. *See also* Mothers Against Drunk Driving.

Suggestive selling A merchandising technique used in sales; involves mentioning items or features that a customer has not ordered.

Sushi A Japanese dish combining cold rice, raw fish, and vegetables.

Sweet acidophilus A pasteurized milk to which *Lactobacillus acidophilus* bacteria have been added. This type of milk can be consumed by people who lack the enzyme lactase. *See also* Lactase; *Lactobacillus acidophilus.*

Sweetened condensed milk A sterilized milk with 40 percent added sugar and half of its water removed; used mainly to prepare baking products, desserts, and dessert toppings.

Table d'hôte A type of menu that offers a complete meal at a fixed price and at specific times. *See also* À la carte.

Table service The placement of food on a table and the style in which it is done. *See also* American service; Buffet service; Family style service; French service; Russian service.

Tabletop A term that refers to the implements and decorations presented on a dining table including the place setting, linens, and condiments.

Take-out service A service style that provides customers with prepared foods for at-home consumption. *See also* Banquet service; Cafeteria service; Family style service; Fast-food service; Full service.

Tankard A tall, one-handled drinking vessel. Early colonial settlers brought silver or pewter tankards with them from England.

Target market The group of customers who would be most likely to buy a product or service over a period of time. Identifying target markets helps entrepreneurs determine needs for new products or services.

Tavern An establishment where alcoholic beverages are served.

Theme or concept restaurant A restaurant that is centered around entertainment or decor. Menus in these restaurants typically include illustrations that reflect the restaurant's theme or concept.

TIPS *See* Training for Intervention Procedures.

Toxins Poisonous substances that are products of living organisms.

Training for Intervention Procedures (TIPS) A training program for servers that helps them to identify when a guest is in danger of becoming intoxicated.

Trenchers Thick pieces of round, dried bread that were used as plates during the Middle Ages in Europe. They absorbed gravies and sauces, and diners often ate them at the end of a meal.

Trend A current style or preference. Identifying trends is an important part of developing new products and services.

Trichinella spiralis A microscopic, worm-like parasite frequently found in pork products. This parasite can be transmitted to humans and cause trichinosis, a parasitic disease that can be avoided by cooking pork to a well-done stage.

Truffles Edible, potato-shaped fungi that grow underground and are considered a great delicacy.

Typeface The style of printing type that is used in a menu or other printed piece. Different typefaces can be used on a menu to create a pleasing overall design.

United States Department of Agriculture (USDA) A government agency that enforces state and federal food service regulations. The USDA inspects meat, poultry, and other processed agricultural commodities and foods. It also grades these commodities for quality.

United States Department of Health and Human Services (USDHHS) A government agency that enforces state and federal food service regulations.

Unstructured interview An interview in which broad, unplanned questions are asked. This type of interview is most useful when personal characteristics are important parts of a job description. *See also* Semistructured interview; Structured interview.

USDA *See* United States Department of Agriculture.

USDHHS *See* United States Department of Health and Human Services.

Vichyssoise A potato and leek soup made with chicken stock and mixed with chilled cream.

Volt A unit of electrical potential or potential difference.

Volume feeding The act of providing food for large numbers of people.

Vomatorium An infamous feature of Roman architecture that provided guests a facility where they could vomit in order to return to the table and gorge themselves again.

Wages Money earned by the hour. *See also* Salary.

Wait staff The employees who provide for the service needs of guests at full-service restaurants. Often called servers.

Walk-in freezer A thermally insulated room that maintains subfreezing temperatures for the rapid freezing and storage of perishable foods.

Walk-in refrigerator A thermally insulated room that reduces and maintains temperatures appropriate for the storage of perishable foods.

Wattage An amount of power expressed in watts, or the amount of energy each piece of equipment needs to work.

Weekly bid A method of purchasing that consists of buying over the telephone, followed by a written confirmation after the bid is accepted. Also called weekly quotation, this method is commonly used to purchase items that need to be ordered in small quantities, such as fresh produce. *See also* Fixed bid.

Whipping A mixing technique that involves beating ingredients together at high speed with a mixing machine. This process incorporates air into a substance and makes it light and firm.

Whitewash A thin mixture used to thicken gravies and soups. It is made with cold water, flour, and milk or stock, which are blended well to a thin, creamy consistency without visible lumps. Whitewash is sometimes called slurry.

Wholesale cuts The large sections of meat into which a carcass is divided and which are inspected and stamped if they pass inspection for wholesomeness.

Wholesomeness In meat inspection, USDA approval for wholesomeness guarantees that purchased meat is from healthy animals slaughtered and processed under sanitary conditions.

Wine steward ***(sommelier)*** A restaurant employee who specializes in fine wines and knows which wines will complement different types of food, sauces, and spices.

Worker's Compensation Act An act that provides compensation to workers injured on the job. Employers must provide income and medical benefits to employees injured in work-related accidents, and they must provide safety and accident prevention training programs and encourage safety in the workplace.

Yield A specified number of portions that a recipe will produce.

Yield grade A number from 1 to 5 assigned by a USDA inspector to indicate the amount of lean, edible meat that a carcass contains. Yield grade 1 is the same as USDA Prime, and yield grade 5 is the same as USDA Commercial.

Index

T

U

V

W

Y

Z

PHOTO ACKNOWLEDGMENTS

For permission to reproduce the photographs on the pages indicated, acknowledgment is made to the following:

Part 1 p. 4 THE BETTMAN ARCHIVE; p. 8 compliments of THE GOLDEN LAMB, Lebanon, Ohio; p. 9 photo courtesy of WALDORF-ASTORIA; p. 10 The Delta Queen Steamboat Co., New Orleans, La.; p. 15 (top) © Richard Anderson; p. 15 (bottom) Barth Falkenbert/STOCK, BOSTON; p. 17 photo courtesy of The Manchester Inn, Middletown, Ohio; p. 18 Holiday Inn, Inc.; p. 19 photo courtesy of SEA ISLAND, The Cloister, Sea Island, Ga.; p. 20 © Michael Wilson; p. 22 (top) photo courtesy of the Institute of Food Technologists; p. 22 (bottom) photo courtesy of KIMBERLY-CLARK CORP., Services & Industrial Sector; p. 23 A. Blake Gardner III, photographer.

Part 2 p. 29 photo courtesy of Hewlett-Packard Company; p. 32 J. Berndt/STOCK, BOSTON; p. 33 3401 CAFE FOOD COURT, THE SHOPS AT PENN, UNIVERSITY OF PENNSYLVANIA; p. 34 Ponderosa Steakhouses, a division of Metromedia Steakhouses, Inc.; p. 35 photo courtesy of ARIS MANUFACTURING COMPANY, a division of Metromedia Steakhouses, Inc.; p. 36 photo courtesy of ARIS MANUFACTURING COMPANY, a division of Metromedia Steakhouses, Inc.; p. 47 courtesy of World Tableware International; p. 55 HERSHEY FOOD CORPORATION; p. 63 Robert E. Daemmrich/Tony Stone Worldwide; p. 65 sample courtesy of Remanco Systems, Inc.; p. 71 photo courtesy of SEA ISLAND, The Cloister, Sea Island, Ga.

Part 3 p. 78 photo courtesy of CENTERS FOR DISEASE CONTROL; p. 81 photo courtesy of NOR-LAKE, Hudson, Wis.; p. 95 photo courtesy of The CHEF'S CATALOG, Northbrook, Ill.

Part 4 p. 115 photo courtesy of the restaurant at The Phoenix, Cincinnati, Ohio; p. 124 courtesy of Hobart Corporation; p. 125 courtesy of Hobart Corporation; p. 127 courtesy of Hobart Corporation; p. 129 courtesy of Cambro Manufacturing Company; p. 134 courtesy of Cambro Manufacturing Company; p. 135 courtesy of Cambro Manufacturing Company.

Part 5 p. 143 from AUDREY'S A PRACTICUM FACILITY AND RESTAURANT OF JOHNSON & WALES UNI-

VERSITY; p. 144 courtesy of SHERATON UNIVERSITY; p. 168 photo courtesy of FLORIDA TOMATO COMMITTEE.

Part 6 p. 182 photo courtesy of BK's NEW ASIA, West Chester, Ohio; p. 195 USDA-SCS photo by Donald C. Schuhart; p. 196 photo courtesy of the NATIONAL LIVESTOCK & MEAT BOARD; p. 198 Courtesy of the NATIONAL PORK PRODUCERS' COUNCIL; p. 199 USDA photo; p. 200 photo courtesy of the National Broiler Council; p. 201 photograph by A. Blake Gardner III at The Culinary Institute of America; p. 241 American Dairy Association; p. 246 American Dairy Association.

Part 7 p. 258 used with permission from McDonald's Corporation; p. 262 (top) Handy Soft Shell Crawfish, Baton Rouge, La.; p. 262 (bottom) Handy Soft Shell Crawfish, Baton Rouge, La.; p. 285 photo courtesy of NCR Corporation; p. 286 photo courtesy of NCR Corporation; p. 303 (right) used with permission from the McDonald's Corporation; p. 311 Jeff Greenberg; p. 317 CARROW'S RESTAURANTS, INC., A SUBSIDIARY OF RESTAURANT ENTERPRISES GROUP, INC.; p. 318 CARROW'S RESTAURANTS, INC., A SUBSIDIARY OF RESTAURANT ENTERPRISES GROUP, INC.; p. 322 Dining With Heart Logo courtesy of Thomas Jefferson University Hospital.